ECOLOGY, CIVIL SOCIETY AND THE INFORMAL ECONOMY IN NORTH WEST TANZANIA

The Making of Modern Africa

Series Editors: Abebe Zegeye and John Higginson

Community Health Needs in South Africa
Ntombenhle Protasia Khoti Torkington

Consolidation of Democracy in Africa
A view from the South
Edited by Hussein Solomon and Ian Liebenberg

Ghana in Search of Development
The challenge of governance, economic management and institution building
Dan-Bright S. Dzorgbo

Regional and Local Economic Development in South Africa
The experience of the Eastern Cape
Etienne Louis Nel

Agrarian Economy, State and Society in Contemporary Tanzania
Edited by Peter G. Forster and Sam Maghimbi

Entrepreneurial Ethics and Trust
Cultural foundations and networks in the Nigerian plastic industry
Yakubu Zakaria

Growth or Stagnation? South Africa heading for the year 2000
Mats Lundahl

Sudan's Predicament
Civil war, displacement and ecological degradation
Edited by Girma Kebbede

Ecology, Civil Society and the Informal Economy in North West Tanzania

CHARLES DAVID SMITH

Ashgate

Published by
Ashgate Publishing Limited
Gower House
Croft Road
Aldershot
Hampshire GU11 3HR
England

Ashgate Publishing Company
131 Main Street
Burlington, VT 05401-5600 USA

Ashgate website: http://www.ashgate.com

British Library Cataloguing in Publication Data
Smith, Charles David
Ecology, civil society and the informal economy in north west Tanzania. - (The making of modern Africa)
1. Sustainable development - Tanzania 2. Tanzania - Economic conditions - 1964- 3. Tanzania - Social conditions - 1964-
I. Title
338.9'678

Library of Congress Control Number: 99-073321

ISBN 0 7546 1068 3

Reprinted 2002

Printed in Great Britain by Biddles Limited, Guildford and King's Lynn

Contents

Acknowledgements

This project was supported by a Postdoctoral Fellowship (1986-1988) and a research grant (1988-1990) both provided by the SSHRC, the Social Sciences and Humanities Research Council of Canada; as well as a travel grant from the Centre for Refugee Studies, of York University to cover the Rwandan crisis of 1994.

Many people helped at various stages including anonymous reviewers. Harold Wolpe, Maxine Molyneux and Michael Cowen provided comments and suggestions on earlier but related PhD dissertation work. Warwick Armstrong, Samuel Noumoff, and Rosalind Boyd, all Directors of the Centre for Developing Area Studies of McGill University, at different times provided much needed logistic support as well as intellectual guidance. Christopher Lwoga of the Department of Sociology, University of Dar es Salaam helped with research clearance and logistic support. Magdelena Ngaiza and Anna Tibaijuka also at the University of Dar es Salaam provided library resources as well as allowing me access to their publications and manuscripts. Gration Kabalulu and Jackson Bubewa provided field assistance and cultural direction and both became friends as well as assistants. Lesley Stevens was my companion and associate in the first nine months of fieldwork as well as editorial assistant. I am extremely grateful that she helped me through a bad bout of malaria before fieldwork began. Naushad Gulam and Peter Shmidt assisted with transport during the fieldwork. Daria Stermac worked together with me on the York University proposal but was unable to get out to the refugee camps because of poor health. Goran Hyden wrote a very supportive Foreword and Sherry Olson, Department of Geography at McGill provided very detailed and useful comments on an earlier version of the manuscript. I thank all colleagues who I met at conferences and other venues for their comments and ideas, especially James Graham for his continuing encouragement and support. Most of all I thank the Haya villagers who let me take up their valuable time and were extremely helpful and hospitable to a sometimes culturally naive guest. Above all I hope this project will help these villagers in some ways, even though very few of them will get a chance to read any of this text.

Foreword

When independent African states came into being some thirty years ago, the world was bubbling with faith in progress. Under the influence of technological, social, and economic change, development began to be seen as a matter of transition from tradition to modernity. Modernization was viewed as a continuous series of changes accompanying the acquisition of knowledge and its effects on ways of getting things done.

The result was that development strategies called upon Africa to adopt contemporary models with historically alien antecedents. Pet notions both from the East and the West were applied on the assumptions that they would help Africa into the global mainstreams of progress. As a result, the last three decades have been a period of endless and sometimes shameless experimentation in Africa. Governments and donors have tended to ignore the narrow margins of survival that characterize African countries at all levels. Above all, they have failed to look adequately for African solutions to African problems.

Public institutions in Africa have run on imported energy rather than domestically available resources. Development on the continent, therefore, has been a process in which words and numbers bear little relationship to its material and social realities. Africa has been brought to adulthood with little understanding of and respect for its own dynamics and abilities. It is no wonder that it has gone down rather than up in the last decade.

It is clear that the earlier assumptions about Africa being a *tabula rasa* is wrong. The institution of development cannot be treated in isolation from their social context and must be fitted into the material and cultural dynamics of each individual country. "Modern" and "traditional" values may be at loggerheads with each other, and the latter may sometimes impose restrictions on development. Yet traditional values are enabling as often as they are impeding. They are certainly not stagnant but constantly developing in response to a changing environment.

In this volume, Charles Smith is demonstrating this through a very careful and detailed study of the farming systems practised by a differential set of farmers in Bukoba District, Kagera Region in Northwest Tanzania. Smith spent almost three years in the area and has assembled one of the finest micro studies of what is happening at the local level in Tanzania, reminding at least myself of the extremely detailed but useful work in the same region by Jorgen Rald in the late 1960s. It has become more and more difficult to carry out such detailed micro studies in recent years and it is therefore that this study takes on special importance.

But Smith's study is significant also for another reason. Far too many studies on Africa have tended to generalize conclusions not only from one country to another but also across social categories. This volume is different. It demonstrates the great variety that exists in a local area and points to the growing social stratification that is taking place in the light of economic liberalization policies.

What is particularly remarkable about Smith himself is his firm belief in the possibilities of grassroots development. His study provides ample evidence that local values and institutions are alive and very much the determinants of social actions. In this respect, his empirical evidence is relevant not only to current theorization in the social sciences but also practical development work.

To be sure, I find myself sometimes overwhelmed by his optimism, but it is important that those who have had occasion to study the vibrancy of Africa's local institutions make a point of highlighting it since so much of what is being written on Africa today carries a "doomsday" message. I also find myself sometimes in disagreement with Smith about his interpretation of others' work. For example, in my own book, *Beyond Ujamaa in Tanzania* (1980), that he cities often, I do not really argue that peasants are subordinated. Such a position is more closely associated with the orthodox neo-Marxists who analysed the Tanzanian countryside in class terms.

Such disagreements, however, do not conceal the tremendous value of this volume. Even if it deals with only one small area of Africa, it does so with commitment and intellectual creativity in ways that make it a pleasure to read; a deep and penetrating analysis that is far too rare at least in the political science field.

Goran Hyden

1 Introduction

"Afro-pessimism" dominates the public image in the developed countries where Africa is synonymous with apocalyptic visions of war, corrupt autocratic leaders, AIDS and other diseases, poverty, drought, famine and environmental disasters. A Southam Press "News Graphic" presented a map of the countries of Africa marked with a four symbol legend: "drought within the last ten years" (seventeen markings); "civil war and armed conflicts" (fourteen markings); "AIDS epidemic" (nine markings); and "famine" (thirteen markings). The Western mass media have not been overly excited by recent attempts to implement multi-party democracies, however the academic establishment have turned liberal reform into the scholarly equivalent of a growth industry. For example the *African Studies Review,* is the journal of the largest African studies association in the world; it publishes articles on history, culture, theory, art, literature etc. In the 1994-1996 volumes (nine issues) there were forty-six published papers, of these twenty fall into the category of what Dickson Eyoh calls "the new political sociology for Africa". In his article in volume 39 (3) Eyoh quotes over two hundred references the vast majority of them published in the past decade. Similarly all the major journals are heavily into political analysis in this period of democratization, and the publishing lists are full of books and readers that fall under the rubric of political sociology. In much of the literature, poor governance and corruption are the chief villains that perpetuate poverty. Afro-pessimism is therefore not only a mass media phenomenon but more a rule than an exception in the scholarly African Studies literature as well.

Africa merits little mass attention and only then as a continent in crisis. A typical example is a syndicated essay by William Pfaff in *The International Herald Tribune,* the by-line challenges "Africa: Can a formula for Stability and Progress be Found?" Nobody denies the enormity and severity of Africa's crisis situations: but those who search for a more detailed and balanced portrait must recognize the immensity and diversity of a continent with over fifty nations and more than eight hundred ethnic groups. It is unwise to generalize: yet few popular commentators have either the time or the motivation to carry out careful fieldwork. Regional and national case studies, and theoretical analyses can provide an alternative to simplistic negative stereotypes, for example, this book examines only one region and one ethnic group, and even at that, it is a partial picture of a complex reality.

Why are concepts such as democratization. civil society, voluntary associations, the civic realm, elections, national conferences and related events and structures so fascinating to those who do have the time and motivation to do careful research? Why is there a proliferation of articles and books on these themes? Goran Hyden expressed the dilemma in his Presidential address to The African Studies Association, "efforts to build a new democratic political order in [African] circumstances are not easy". However at least there is a glimmer of hope that political reforms will lead to greater freedom as well as material well-being.

This work tries to avoid Afro-pessimism however it does not fall squarely into "the new political sociology for Africa" but rather tries to be as realistic and objective as possible. The fact that I am resident in East Africa and have spent eight years here does not necessarily give me a great advantage. Most Africanist scholarship is published outside the continent and few scholarly and scientific journals are available at the university where I teach. In fact the only chance I get to catch up on scholarly research is on visits to North America. At the beginning of this project, before I went to Tanzania to carry out this research, I read almost every work published on the Kagera region, as well as many archival documents from colonial times, 1900-1961. After my PhD thesis on colonialism had been accepted in 1985 I considered myself well prepared to tackle the modern era. Nevertheless, four years in Tanzania convinced me that everyday life for the majority of people were based on realities which few researchers have written about. Almost all the literature on the Kagera Region assumes that the average farm household produces coffee as a cash crop and lives on home-grown subsistence crops. This is entirely consistent with the "facts": coffee is Tanzania's principal export crop, the region is one of the largest coffee producers, official statistics are based on official crop marketing, and the Kagera Cooperative Union kept a careful record of all its coffee purchases and sales.

But in the real world of Kagera villages coffee sales only made up about one-sixth of the actual monetary income of my informants. Coffee sales and all other officially registered cash transactions are in reality only one small portion of the cash-based village market place. The second major discrepancy between reality and the literature was the insufficient attention paid to about one-quarter of the region's households headed by women. The profile of these farms is so distinctive that it significantly changes and in fact challenges the notion of the average farm household. Women headed households earn a living by intensively exploiting fewer resources and they contrast markedly with male-headed households who control more land mainly from inheritance under customary law.

I hope this book will be of interest not only to Tanzania specialists but more generally to students and the concerned public who want to look beyond the negative Africa stereotypes and to search out more concrete details, as well as underlying structures. The major difference between this book and most of the other literature on Tanzania (especially material published before the mid 1980s) is its focus on the underground social/ economic realities and its indications of positive trends. Trends which, in my view, provide a model of sustainable development that will have a much wider applicability than the specific geographic area and its regional social/ cultural/ political/ economic context.

Themes and Concepts

Africa must somehow meet the challenge of how to feed a growing population within ecosystems pressed to their limits and in some instances degraded by irreversible processes such as loss of valuable topsoil through laterization and erosion. Food security in the 2000s and beyond will require greater productivity, alternatives to cash crops, larger incomes with which to purchase food, an end to environmental degradation, and of course population control.

Attempts at planning for sustainable development are handicapped by inadequate research, confusing and inconsistent terminology, and competing conceptualizations of "correct" development planning. It is both difficult and costly to collect information about the informal economy and hidden resources: hidden resources include material goods, such as unharvested cassava, as well as the energy, skills and resourcefulness of Africans.

In the chapters that follow I will present data obtained over ten years of field work in the Kagera Region of Tanzania: in 1994 the Kagera River became notorious because it is the boundary between Tanzania and Rwanda. During the genocide or 1994 so many bodies were dredged from the river that many people in the region stopped eating fish from Lake Victoria fearing contamination. The genocide and the incursion of a half million refugees was a temporary occurrence, however the long-term consequences are not yet known. The acute problems of militarization, ethnic conflict, economic disaster, and civil war did not cause a similar crisis in Tanzania, however the spill over from Rwanda is causing a geo-political crisis in the Great Lakes Region that will continue to heighten tensions into this millennia as Rwandan and Uganda troops as well as armed Hutu militants patrol the frontier areas of the DRC, Democratic Republic of the Congo, threatening the unstable regime of Laurent Kabila and his various allies.

The impact of forced migration on the smallholder farmers of this study indirectly exacerbated a cumulative series of pressing problems: insects threaten harvests of the plantain crop (the staple regional food); cattle overgrazing causes degradation of pasture land; the division of labour on gender lines underutilizes available male work potential and undervalues traditional female tasks, and forced migration reduced available timber and grazing land in the border areas.

The chapters of this text all address in one way or another the concepts or problematics which guide development planning. Despite the complexity of the issues and the diversity of terminology, I believe that there are two fundamentally opposed analytical approaches to the problems of the peasantry and economic development, I have chosen to label these "planning futures for Africa" and "small is beautiful".

Development planners who favour the first approach tend to view peasants as an obstacle to development best disposed of, or subordinated, to make way for a more modern social and economic system: those who argue for smallholder innovation generally believe peasants can and do sustain themselves, and need to continue along the path of democratic reforms in order to achieve sustainable development— based on the farmers own knowledge of local conditions. I believe that in Tanzania today economic, social and political conditions favour an approach based on small is beautiful type innovations as the only realistic option.

The key question is how to reverse a trend toward overcrowding and degradation of a farming system that has proven itself to be sustainable over centuries. Smallholder agriculture looks very low tech and therefore traditional, nevertheless, Kagera farmers have succeeded in adapting to their environment by using technically appropriate simple-yet-effective innovations. The challenges for the future are to overcome the constraints of the class and gender system, to encourage the tenacity and resourcefulness of women farmers and to provide meaningful work for relatively idle young males. Increasing coffee production is the best way to generate much needed foreign exchange by exploiting under utilized male labour. The best way to bolster production is for the regional cooperative society, or alternative organizations now that privatization is underway, to offer incentive goods and to do so in a manner that does not threaten the integrated regional system of farming and related enterprise.

Some Haya farmers, such as the Bishubi brothers have already solved the problems of sustainable development. Joni Bishubi is one of the richest farmers out of the two hundred and fifty sampled households. His family farm, in combination with trading and artisanal activities provides him with a relatively large and secure cash income and a surplus of food and other subsistence goods. Few women farmers can attain rich farmer status and enjoy this comfortable lifestyle, those who do usually owe their success to either inheritance or involvement in illegal activities — there are only five cases of successful female farmers and they include a black marketeer and two former prostitutes.

The middle farmer class are holding their own, they are relatively secure except in times of crop failure caused by unusual weather, war, or other tragedies. Most middle farmers have either inherited land or work at salaried employment or trading and have purchased land. Gration Kabalulu is one such farmer, he holds a salaried position as an officer of the Ministry of Youth and Culture in Bukoba town but he also operates a mid-size farm in the "pombe village" described in a later chapter. Constance Bubewa is one of a very small number of women middle status farmers. Her household acreage is too small to provide adequately for her family — this is the most common pattern for female-headed households — but her employment as a teacher in the village school gives her enough income to qualify as a middle farmer.

Most women farmers are relegated to insecurity of income and food because they occupy very small acreage and do not have adequate capital or skills to earn a good living in trade or crafts. The majority of younger Haya leave the farms for a decade or two and work in the towns in order to accumulate farm capital. Youth who stay in the villages, together with women and the elderly debilitated, make up the remainder of the poorest farmers, the ones most at risk from land shortages; a problem compounded by environmental degradation. In the conclusion, I argue that the innovations developed and used by the rich and middle farmers could be used to assist the poorest Haya.

If conditions are to improve (rather than worsen as most analysts predict for tropical Africa) then the Haya must improve crop yields as well as their earnings in trade and artisanal enterprises. Training in practical skills is sorely needed, as are ways of providing capital for purchasing and managing relatively expensive resources such as dairy cattle. If the Haya can manage to control population growth then their farming system and lifestyles can be sustained and supported by the types of innovations already being adopted.

2 Conceptualizing the Peasantry

Haya Tradition and Innovation

Tanzania is a predominantly rural nation; most farmers are poor smallholders; therefore, grassroots innovations provide the best prospects for sustainable development. Haya villagers have adopted many innovations: new crops, new methods of cultivation, trading/hawking, and very small scale artisanal enterprises. Notable innovations include: tree farming for firewood, charcoal and lumber, raising purebred dairy cattle, production of burnt bricks and other construction materials, and, newer methods of brewing banana beer or distilling illegal liquor. The cash nexus is the motor force driving this small-scale commerce, the state neither monitors nor controls nor taxes this informal economy.

These innovations, rooted in Haya customs and ecology, enhance and extend the traditional way of life. Although this chapter discusses issues which appear highly abstract, such as definitions of the peasantry or some of the debates on how best to promote development in peasant dominated societies, however, these abstractions have very material and very practical consequences. If planners view peasants through the lens of modernization theory, "planning futures for Africa", then smallholders personify traditional society which is non-cooperative and non-productive and the preferred policy option will often be to try to eradicate peasant cultures. If on the other hand, planners conceptualize peasants as potential entrepreneurs and innovators then they are more likely to promote a "small is beautiful" more gradual approach to development. In Tanzania where nine-tenths of households are situated on small plots, the state and international agencies have tried and failed in their efforts to "rationalize" peasant production into *ujamaa* villages and experiments with larger-scale farming estates have failed, as this sector diminished to one-fifth of its former scale between 1968 and 1990.

Who are Peasants?

The hallmark of peasant farming is technological simplicity and a relatively non-specialized division of labour; agricultural production is carried out by "small producers with simple technology and equipment, often relying primarily for their subsistence on what they themselves produce" (Firth, 1951, p.84). Peasants usually organize farm labour into household units who decide how to allocate its available work force (Bryceson, 1991).

Most analysts accept the notion that "peasant culture is distinct from, but related to, the larger culture" (Klein, 1980) and that a "folk" or "little" peasant tradition is the antithesis of a "great (industrial) transformation" (cf. Redfield, 1956; Fallers, 1961; Polanyi, 1968). Many Marxist and liberal social scientists hold the "little tradition" responsible for low productivity in African agriculture, since they believe that peasant cultures inevitably promote technical and political backwardness and a pre-scientific world view. The alternative strategy for social planners is the "great transformation" — a "planned future" based on the adoption of large-scale capital intensive enterprises, usually this type of "progress" would require the dispossession of large numbers of smallholders.

I prefer Schumacher's "small is beautiful" model of sustainable development, in which peasants readily adopt innovations which are appropriate to their conditions of life. For most of sub-Saharan Africa the "innovation" approach will work best in the short and medium term provided that conflicts can be resolved between peasant groups — divided by property, ethnicity and gender. The "subordination" of the peasantry may be an inevitable historical transformation, but Africa in the 2000s has few alternatives to peasant agriculture because the industrial sector is usually incapable of supporting its rapidly growing population. The time for Africa's "great transformation" has not yet come.

Both strategies acknowledge the autonomy of the "little tradition" peasants are only peripherally connected to the state and the national and world markets. However, planners following the path of modernization theory usually echo Weber, Parsons, Levy and other pioneers who believed that "traditional society" and peasant economy was the primary structural impediment to development. Talcott Parsons formulated the concept of pattern variables to distinguish modern from traditional societies; patterns embedded in the cultural system include affective versus affective-neutral relationships, a particularistic versus a universalistic world view, ascription versus achievement , functionally diffuse versus functionally specific relationships and a collective versus a self orientation. The bottom line is that the peasant's backward mentality is the principal explanation of why modernization has not yet occurred.

The "small is beautiful" tradition argues that peasant autonomy is perpetuated by national and international forces which benefit from low wages and producer prices. Small farmers are blocked by inadequate resources not their "traditional" world view. Any meaningful attempt to significantly improve productive capacity must arise from and serve the needs of Africa's rural masses: more aid is not the answer but rather more effective international and national assistance.

Is it Necessary to Subordinate the Peasantry?

Planned futures/ modernization arguments are usually based on two related assumptions, firstly, that peasant cultures are organized on the basis of social and moral imperatives and therefore are non-innovative and traditional; secondly, that small-scale agriculture is too inefficient and unproductive to generate the surplus necessary to finance material progress. In the years since Independence in 1961, the Tanzanian Government has tried (usually unsuccessfully) to organize initiatives meant to modernize the rural sector which comprises eighty-six percent of national population.

Goran Hyden's many publications on the subject of Tanzanian peasant backwardness have bolstered his well-deserved reputation for careful research and insightful analysis. According to Hyden, the peasant's "economy of affection" (reminiscent of Parson's affective versus affective-neutral distinction) is a primary factor holding back "the development of science and technology" or "*nature artificielle*". In his schema African agriculture remains "pre-scientific" because:

> What the peasant needs to know as a successful cultivator is not systematized into a set of abstract hypotheses ready to be tested. This knowledge is not 'universalized' into separate theories . . . In this organic environment, there is little understanding and tolerance of experimentation and limited scope for problem-solving of the kind that we associate with an inorganic environment. (Hyden, 1983, p.5)

Throughout this book, I argue that Haya peasants *are* able problem-solvers, who admittedly must operate within circumstances of material poverty, judged by the standards of the industrialized world. Peasant problem-solving is geared to the economic restrictions of peasant life. For example, peasants don't use village-made wooden wheelbarrows because they work better or even last longer than imported metal ones or even because an imported one could not be repaired locally. It doesn't matter to a Haya villager whether or not a wooden wheelbarrow or a metal one will optimize profits because only the wealthiest Haya farmers can afford the imported wheelbarrow.

Development theorists sometimes assume that the peasant is a conservative and traditional consumer, however empirical evidence suggests that peasants have no choice but to spend their small amounts of capital on goods and services that have been proven appropriate to the village environment. This study highlights an important but often neglected fact; whether a smallholder uses centuries old traditions or recent innovations is more often a decision based on wealth and access to capital than any inherent conservatism.

The Haya of the 1990s seem to belie the assumption that the peasants exist in a "moral economy" (Scott, 1976) in which their security depends on "the maintenance of social links with neighbours able to help in times of need" (Richards et al., 1973, p.4). The "moral economy" model assumes that neither long-term subsistence nor social status require much monetary wealth, therefore economic rationality does not necessitate the "profit-maximization calculus of traditional neo-classical economics" (Scott, 1976, p.4). The descriptions of peasant societies by subordination theorists therefore use terminology such as, "risk-averse" (Scott, 1976, p.4) followers of a "consumption imperative" (Hyden, 1987, p.661) or governed by the "law of subsistence" (Hyden, 1986, p.625). Furthermore, this moral economy or economy of affection is the principal block to development. Since peasants are dispersed and unorganized they generally do not directly confront the state but rather employ what Scott calls the "weapons of the weak", they exit from the state and the market and therefore may remain "uncaptured". It is a paradox, but their collective strength rests on geographical dispersion of the peasant household and its limited consumption. Most proponents of modernization/subordination theory believe that as long as peasants control their own production and marketing then their society must remain predominantly "folk/rural/traditional" and will not become "universalistic/urban/modern" (Foster, p.1967, 12; Post, 1971).

How "Moral" is Peasant Economy?

The predominance of small-holder farming is not always synonymous with a "moral economy" or an "economy of affection". Pre-colonial Haya society was less communal than most other Tanzanian ethnic groups: like the other kingdoms of the Great Lakes zone, it was hierarchical and feudal. The kings and the royal clans maintained order and stability by imposing harsh punishments; one elderly informant claimed to have witnessed certain offenders publicly executed by being impaled (this occurred when Tanganyika was officially a German colony). Brutal fiat by an aristocratic clan hardly connotes an economy of affection or moral economy.

Communal and neighbourly ties are certainly important to the Haya, but there are many areas of social life in which sharing with the community has been replaced by individualistic behaviour. The clearest examples were: numerous break-ins to houses and complaints of theft, the reticence of most families to share meals or home-brewed beer with neighbours and clan members, and a decline in gift giving and hospitality at weddings and funerals. Another indication of the breakdown of traditional social obligations is the problem of AIDS orphans. PARTAGE, an NGO from France was caring for 1250 orphans in December 1990 and they estimated a total of 25,000 AIDS orphans in the region. Increasing numbers of orphans are not being adopted by their extended family. No one seems to have very accurate information on the full extent of this problem which may be as serious as in neighbouring Uganda.

Theft is more prevalent in the 1990s than in the 1950s, even though there was probably more real wealth in the 1950s and the British had already abolished most of the cruellest forms of corporal punishment. Nowadays all who cannot defend themselves are at risk and even the poorest may be victimized. To cite one example, during my stay in a small village, an elderly woman living alone was killed by thieves who broke down her door with a large stone and removed her coffee harvest (two gunny sacks of unhulled coffee worth less than fifty dollars). Many of the elders reported that younger people lacked respect for the old traditions and had turned to theft as a way of life. Underemployed landless young men in the village were the ones whose the endurance, strength, and dearth of alternative ways to earn a living made them the most likely to become thieves.

To protect themselves villagers devised their own security measures. Even the poorest households were built with padlocked wooden doors and shutters on the windows: wealthier households used door locks and protected the windows with iron bars set in cement or small glass panels set in metal frames. As an additional security measure, many of the richer farmers and town dwellers kept watchdogs, the more *kali* (fierce) dogs are highly prized.

The most important reaction to widespread theft is the *sungu sungu*, a local vigilante police force organized as a secret society. They patrol in groups of a dozen or so wearing sock masks and burlap sack clothes. In the five villages we studied the *sung sungu* imposed a nine PM curfew and they were known to beat up anyone they caught outside their homes after curfew.

Besides the problem of theft, there are other clear indications of how a cash economy has replaced or is replacing social customs based on sharing and reciprocity. In the past it was customary to share food with guests and neighbours visiting around meal times, whereas now, visitors are very rarely offered food to share. Similarly, home brewed beer and liquor have become commodities generally bought and sold rather than reciprocally exchanged. Because of the large number of AIDS-related deaths, gift giving at funerals has been drastically reduced or curtailed altogether. Another interesting indication of the shift to a generalized money economy is the sale of snacks at weddings where there are large numbers of people congregating. Guests buy samosas and other snacks from enterprising village youth who profit from the fact that contributions from the groom's family and neighbours do not provide enough food for all. Besides having to buy food from vendors, guests also bring their own supplies of banana beer and liquor.

Sons and daughters living in towns did not send back regular cash remittances large enough to ease the financial strains faced by their cash-poor parents. In the sample, the average household had at least one child living away from their family of origin, employed in a large enterprise or running a farm or small business. Nevertheless, sampled household heads received only three percent of their reported cash incomes from regular remittances or irregular gifts.

The "economy of affection" is not the metaphor that I would use to describe the predominant social and economic interconnections. In Northwestern Tanzania, the peasant economy is cash poor and small scale, but the impersonal laws of cash sale based on supply and demand seem to override the moral obligations of custom in most instances.

Small is Unproductive

The subordination argument assumes that the peasant's traditional mentality is the social foundation for a hopelessly unproductive agricultural and industrial base. Hyden, for example, seems to follow Marx's classic formulation, in *Das Capital,* Marx asserted that progress requires "primitive accumulation".

> In the history of primitive accumulation ... the expropriation of the agricultural producer, of the peasant, from the soil is the basis

> of the whole process. (Marx, 1976, p.876)

Hyden indeed echoes this statement:

> History has demonstrated that the development of modern society is inconceivable without the subordination of the peasantry. (Hyden, 1980, p.16)

Hyden disagrees with Wolpe (1975) that the capitalist mode of production obtains net benefits from its exploitation of the peasant mode of production. Hyden (1980, p.17) believes a modern society requires that smallholders be replaced with more efficient forms of production based on economies of scale; it "does not really thrive on the 'restructured' pre-capitalist forms of production".

Hyden and the subordination theorists argue that development will require the radical separation of potential wage workers from the means of production (proletarianization). The logic of subordination dictates that even if the urban sector benefits from small-scale technically backward agriculture, which provides cheap food, child care and other essential goods and services, it could derive even greater benefits from "large capital-intensive estates". Keith Hart, in his study of West African agriculture, proposes that publicly or privately owned "large, capital-intensive estates" are the most effective way to achieve "a long-run dynamic of economic development through labour specialization, capital investment, and productive innovation" (Hart 1982, p.154; Johnston 1986, p.157).

Subordination strategists are correct to argue that in the longer-term agricultural productivity can only be increased by introducing improved technology and replacing some indigenous customs with norms more conducive to capitalist work patterns. But before primitive accumulation and subordination can take place a great deal more intermediate level development as proposed by Goody (1971) must take place. Relatively simple, easily-maintained technology must be developed and adapted to indigenous conditions. For example in Kenya's breadbasket, the Uashin Gishu plateau, conditions favour the use of expensive technology to realize economies of scale. In Western Kenya, even small wheat farmers hire combine harvesters from their richer neighbours because they reduce unit harvesting costs. However in Kagera region and other areas where the staple crop is plantain there is no available technology for harvesting bunches of bananas. There could be an improvement in harvesting techniques if head porterage of fifty kilo bunches was replaced by some form of wheeled devices, but such an innovation would need to be made affordable.

Advantages of Small Scale Production

Current conditions in Tanzania make it difficult to take advantage of economies of scale. Furthermore, there are many examples throughout Africa where small-scale production generates better marginal returns than large-scale alternatives. There is as Johnston claims an "inverse relationship between farm size and output" when "the opportunity cost of farm labour is low because of the lack of off-farm employment opportunities". Under such conditions, the small farmer has an incentive to work hard, acquire skills, exercise initiative and make correct "on-the-spot supervisory decisions" (Johnston 1986, pp.157-159). Since the small farmers have so few off-farm employment opportunities, they must work as hard as necessary to survive and they will adapt appropriate affordable innovations and therefore (if rewarded) they can achieve better results than estates:

> . . . there is abundant evidence that small farm development strategies can be a more economical approach to achieving sector-wide expansion of agricultural production than a large-farm strategy when the opportunity cost of farm labour is low ... When structural transformation and economic development reach a point where the relative availability and prices of capital and labour warrant a shift from labour-using, capital-saving technologies to labour-saving, capital-using technologies, then the economies of farm size become significant and small-farm development strategies cease to be appropriate. (Johnston 1986, p.168)

The important point in the above quotation is that a "structural transformation" can only occur after there has been enough accumulation to finance the expenses of infrastructural development. In theory, a smoothly functioning estate sector may be able to finance these, but in Tanzania today the combination of poor efficiency in large enterprises and potential social disruption preclude this type of structural transformation. The opposite trend is more evident: for example estate sector production of coffee, the nation's most valuable export, has diminished to one-fifth its former level in the past two decades.

Smallholder farming in Tanzania makes both economic and ecological sense. Williams (1976, pp.146-147) demonstrates that Tanzanian peasants can deliver goods at a lower cost than capitalist agriculture, and that in Tanzania, economic assumptions based on the supposed superiority of economies of scale are incorrect. Smallholders have several initial advantages over estates: they know the local environment and the real cost of inputs including opportunity costs; they pay their own recurrent costs; they produce goods using lower overheads than estates who must pay salaries, housing, administration and social costs; the smallholder actually has a higher average and marginal rate of savings and investment (to ensure his livelihood) than the estate farmer. The peasant who owns the product has a greater incentive to work than a salaried employee. Small farmers use complementary resources such as intercropping to improve soil productivity and lower labour costs. Peasants always try to maintain soil fertility while capitalist farmers may, destroy prime agricultural land by "mining" it for short-term gains; see Raikes (1986, p.35) discussion of Ismani district. Peasant cultures often exist in a symbiotic relationship to one another as in Lindstrom's (1986) account of barter between agriculturalists and pastoralists in Iramba district.

Innovations at the Grass Roots

In Tanzania, the most successful projects start with grass roots efforts but these initiatives invariably break-down when expanded to the national level. For example, the Ruvuma Development Association settlements were very successful until they were taken over by Tanzania's national villagization programme (Williams, 1976, p.138; Coulson, 1982). Despite registering 8299 villages and fourteen million people by 1979 (Mapolu, 1986, p. 119) the massive drive to create "cooperative villages" has been a costly failure, leading to declining national food and cash crop production.

Even during the 1980s, a period of crisis, Tanzanian smallholders were often innovators. Women farmers, at least those who have secure land rights and can reap the benefit of their labour, successfully cultivate tea in the southern highlands (Odgaard, 1986). Small farmers in the Livigstonia mountains, created Tanzania's breadbasket: they increased grain production sixfold in fifteen years (Rasmussen, 1986) by adapting the most affordable parts of a green revolution package. Despite little official assistance, there was a boom in the use of ox-ploughs between 1975 and 1985: during the same time frame the heavily subsidized tractor programs have completely broken down (Kjaerby, 1986). Masai pastoralists switched from cattle to goats in response to the loss of control over their grazing lands (Arhem, 1986). Due to the high cost and scarcity of Western medicines, the market for folk medicines is flourishing in Tanzania (Heggenhougen, 1986, pp.312-313). Small craft industries (often set up on smallholder plots) are better able to turn a profit and withstand the crisis than the heavily subsidized Small Industries Development Organization (Havnevik, 1986, pp. 287-288).

Maliyamkono and Bachwa, 1990, provide the most comprehensive analysis of Tanzania's "second economy", documenting (in 41 pages of tables) subsistence farming, small-scale unofficial trading, and a host of other economic activities not monitored by the state. Tanzania's "second economy" was the first economy for my sampled households: officially-recognized trading and salaried employment only produced one-fifth of their total cash incomes.

Modernization theory assumes that peasant enterprises (such as the production of herbal remedies or local wooden wheelbarrows) develop as expedient and cheaper substitutes for superior manufactured goods. However even if this is true, peasants are not likely to become proletarianized by market forces in the near future because Hyden's assessment of the peasant relationship with the state is well founded;

"small is powerful".

Political Autonomy and Subordination

The Tanzanian state is not in a position to be able to subordinate the peasantry. The failure of the massive villagization program, is the clearest example of the limits of state power vis a vis the rural masses. Von Freyhold (1979, pp.22-33) believes that in theory the communalization of agriculture in Tanzania could have enhanced peasant forces of production and led to an "intermediate rural transformation". The villagers she studied, in the early 1970s, in the Handeni District, had a good idea of how (in theory) cooperation could raise aggregate incomes (Von Freyhold, 1979, pp.74-75). But because the Tanzanian social formation was not ready for communalization and socialist transformation, the Handeni peasants did not cooperate (Von Freyhold 1979, pp.31, 88-89, 107, 191).

The state issued series after series of directives but failed to complete many of its promised development initiatives (Von Freyhold, 1979, p.191). The three most crucial mistakes made by bureaucrats were, to "coerce", "neglect", and "pamper" ujamaa villagers (Von Freyhold, 1979, p.184). The official cadres and local elites were very poor models, there were many instances of mismanagement and corruption (Von Freyhold, 1981). The main lesson to be learned from this whole process was that the state could move millions of smallholders but it couldn't force them to produce target amounts of specific crops by a directives approach.

As Klein (1980, p.11) points out the state exerts its primary control over peasants by collecting taxes. Yet in Tanzania neither the central government nor the regional government can collect graduated income taxes from the peasantry. The regional government under the banner of "decentralization" has merely replicated the colonial hut and poll tax system by collecting a set fee (the development tax *kodi ya maendeleo*) from each adult.

Between 1982 and 1987 inclusive, the Tanzanian Central Government's total expenditure was 154 billion shillings. Only 16.3

percent of this revenue was raised from income tax: 15.5 percent came from domestic borrowing, 17.5 percent came from foreign loans and grants, 29.2 percent from sales tax, and 21.5 percent from other indirect taxes (see Tanzanian Economic Trends 1988, 52 table 7). In fact the percentage of government revenue from income taxes changed little between 1967/8 when it was 18.7 percent, and 1976/7 when it was 21.1 percent (Coulson 1982, p.193).

The Government of Tanzania's overall revenues (limited in any case because of low levels of national production) cannot effectively tap nine-tenths of the population — the smallholder farmers who do not pay direct income tax. Reliance on indirect taxation (such as sales tax and taxes on imports and export crops) indicates that it is impossible for the government to monitor rural incomes or to devise a mechanism of collecting more direct taxes from the rich and middle peasantry who pay the same tax as the poor, despite the fact that they earn much larger incomes. Even goods and services taxes cannot close the gap, because most of the regional economic activity takes place in the informal sector.

The liberalization drive of the 1990s deregulated National Marketing Boards and the networks of Regional Trading Corporations, but in practice most trading was conducted in the so-called parallel markets even before liberalization and even after the withdrawal of parastatal monopolies, small traders are difficult to regulate and tax. The attempt to outlaw private trading and cooperatives during the 1976-82 period was counterproductive because delayed payments and low prices drove farmers even further away from the official marketing boards as government revenues declined, food imports increased and the parallel markets strengthened (Boesen et al., 1986). If the state is not capable of subordinating the peasants through direct and indirect taxes, what other economic sanctions might it employ? The "grow more coffee campaign" is not likely to mobilize the peasants unless it provides incentives (based on improving real incomes).

I believe that the reported levels of economic decline and stagnation are exaggerated because there has been a shift of resources from the official to the informal economy where most economic activity is taking place. Maliyomkono and Bachwa (1990) accumulated an impressive amount of information to show exactly how most Tanzanians are able to survive because of the informal sector. The best visible evidence for this "second economy" is in the market place. Localized private trading provides most of the goods and services purchased by my sample population, such as: food, furniture, construction materials, and farm tools. Even imports are distributed by a host of small, usually unlicensed, private traders who distribute mass goods (such as used clothing) as well as elite goods (such as reconditioned Japanese pickups).

3 Civil Society and Governance

President Benjamin Mkapa elected in 1995 is continuing along the path taken by Tanzania's second President Ali Hassan Mwinyi. President Mwinyi, under pressure from the IMF and World Bank, initiated "structural adjustment" to transform the system of *ujamaa* or "African socialism" implemented by Tanzania's first President Julius Nyerere. The underlying assumptions of the theory of structural adjustment in Tanzania is that economic liberalization will create an "enabling environment" (Sandbrook, 1995) and usually this is tied to the notion that a national bourgeoisie will arise within civil society and be the dynamic social element that promotes economic development.

Civil Society

Varying notions of the concept "civil society" pervade the history of Western political philosophy. Harbeson (1994: 15-21) provides an excellent summary and review of definitional elements used by pioneer theorists such as Locke, Hobbes, Rousseau, Paine, de Toqueville, Hegel, Marx, Engels and others. Eyoh (1996) in his impressive review of over two hundred recent publications on democratic reform, calls "civil society" the "master concept" in "the new political sociology for Africa." However this master concept is the source of much confusion and difficulty because of the inconsistent way it is theorized and operationalized in the literature on democratization in Africa in the 1990s. I believe that the most useful operational concepts for the analysis of contemporary civil society stem from the works of Max Weber, Antonio Gramsci and a tendency I have chosen to label "resistance theory".

The prevailing tendency in the "new political sociology" is to view civil society as part of a process of democratic reform that is supposed to be based on pluralism and polyarchy (Dahl, 1971; Diamond, Linz and Lipset, 1988). As Robinson (1994) points out polyarchy, liberal democracy and good governance are often conceptualized as preconditions for each other. Good governance requires Weber's ideal type of bureaucracy, one in which the civic realm Hyden (1980 & 1983) dominates to create an "enabling environment" where government power is limited, where individual rights and freedoms are protected, where negotiation and bargaining are the tools used to resolve differences and build national consensus — see for example Doro (1996). These concepts are sometimes given a practical dimension when they are operationalized in constructions such as "the Quality of Democracy Index" formulated by Joseph (1991) and used by the Carter Center, Atlanta.

In principle a technocratic, state-centred, "bureaucratic authoritarian" regime might be able to foster enough transparency to reduce corruption and clientage and promote economic growth, for example see (O'Donnell, 1978; Evans, 1985; or Gold, 1986). However corporatism is rare in Africa due to the "absence of bureaucratic institutions that function authoritatively", see Bratton and Van de Walle 1994, p.458. When forms of corporatism do appear, for example military dictatorship in Burkina Faso, they do not become development states and thus fail to resolve the pervasive crises of legitimacy that carried them to power, Bratton and Rothschild 1992, p.277.

Since the publication of the "Berg Report" World Bank (1981) there has been a very strong tendency to blame most of Africa's economic woes on poor governance, Eyoh (1996). Poor governance in turn tends to be a function of "neo-patrimonialism" and the inability of the state to separate itself from societal influences (see for example Hyden, 1994; Sandbrook, 1995; Callaghy 1987; Nyang'oro, 1989). A system based on polyarchy and pluralism seems the best way of creating civic accountability and transparency at least within the prevailing logic of the "new political sociology for Africa". Beckman (1988) attributes economic stagnation to the *absence* of bureaucratic structures that follow impersonal norms and rules. Terminology such as clientage, neo-patrimonialism, patronage, rent-seeking and predatory behaviour, corruption/ grand corruption, mismanagement and so on, pervade the literature as do metaphors such as "eating" (Schatzberg, 1993) or the "politics of the belly" (Bayart, 1993).

Dahl (1971) and other proponents of pluralism tend to associate polyarchy with what Robinson (1994) calls, *bourgeois liberalism*. This package consists of an open market economy combined with an elected parliamentary democracy, and a national bourgeoisie that is supposed to serve as the historical agent of economic development (cf. Beckman, 1988; Moore, 1966). This model is becoming the "neo-liberal orthodoxy" of the international financial institutions or IFI (Sandbrook, 1994). Callaghy beautifully sums up the logic of this system;

> The presumption of the mutually reinforcing character of political and economic reform lies in the extension of neoclassical economic logic as follows: Economic liberalization creates sustained growth; growth produces winners; winners organize to defend their new found wealth and create socio-political conditions to support continued economic reforms. (Callaghy 1994, p.243)

However the numbers of such "winners" are too few to predict the outcome of economic and political reforms under the prevailing conditions.

The creation of fully bourgeois national elites will require means of defending "new found wealth" from rent-seeking and predatory elites and restructuring "clan politics" and the "economy of affection". Hyden, (1983), Bratton (1989) and others argue that a revamped civil society may be the means of promoting transparency and therefore a prerequisite to replacing the existing predatory elites with more accountable leaders arising from the mainly urban middle class. It is they who possess the new leadership and the organizational and management skills needed in pluralistic societies and who have at least a relative autonomy from the predatory classes.

> Middle classes are the main protagonists of civil society. Middle class elements are prominent in founding and leading civic organizations and in articulating 'universal' values as a means of

> countering particularistic loyalties and building broad multi-class political coalitions. (Bratton, 1994, p.58)

Note the use of Parson's terminology, "pattern variables" include both modern categories such as "universalism" and traditional ones such as "particularism".

At this point it is not clear if it will be possible to replace the existing predatory elites with more accountable leaders arising from the mainly urban middle class: new leadership who possess the organizational and management skills needed in pluralistic societies and who have at least relative autonomy from the predatory classes. There is little hard evidence to support the view that bourgeois liberal democracies will transform Africa.

The Lipset thesis, that economic growth correlates with democracy because economic take-off requires unfettered action by entrepreneurs who develop into a bourgeoisie, (Lipset, 1968; Lipset, Seong and Torres, 1990) is also questionable in the African context. Some analysts conceptualize civil society as tied to the market economy and therefore economic liberalization: however the problem remains how to separate corrupt crony capitalists tied to the state from a growing economy (Crook, 1990; Cohen and Amalo, 1992).

There are many inherent difficulties with the notion that democratization leads to economic growth. As Huntington (1984) and Lipset (1968) point out, too rapid economic growth may lead to rising social expectations and instability; however this is not the case in Botswana, a country with over a decade of economic growth at over ten percent annually (Holm, Molutsi and Somolekae, 1996). Hyden (1994) examines the Lipset thesis in Tanzania and Kenya and comes to the conclusion that Kenya's economic hegemony in the region arose under more authoritarian conditions than prevailed in Tanzania, where a more open and legitimate political system stagnated economically.

Because the neo-patrimonial rent-seeking states are both inefficient and mired in authoritarianism there is a pressing *need* for alternatives to predatory elites and it is tempting to earmark "private oppositional spheres" and assign them a privileged position in civil society, Fatton, 1992, p. 41.

Eyoh (1996) claims the new political sociology for Africa model gauges success, or the lack of it, by teleological ideal type notions. The crucial determinant of what Harvard's Professor Putnam calls "civic awareness" and "social capital" are the norms that both associations and government bureaucracies will be ruled by performance and achievement criteria based on transparency and accountability, moreover pluralized associations shall operate under conditions of civil tolerance and hold each other and the state in check. But these assumptions do not apply in most of sub-Saharan Africa where civil society itself tends to lack civic awareness and social capital (see Putnam, 1995). Tripp (1994) gives an excellent example of how the norms of accountability and civic responsibility were violated in an association supposed to represent disabled women in Uganda. As one member stated "Our leader was unfair and gained a lot from us ... she offered herself the aid meant for us". This quote signals the primary structural dilemma: the "rhyzomic system" permeates all levels of society. Clients and sometimes their brokers, lobbyists and middlemen operate within a system that Olivier de Sardan (1999) characterizes as "a banalisation of corrupt practises" based on a "logic of solidarity" and a "logic of gift giving" that have deep and widespread African roots.

African civil societies are still embryonic in the Weberian sense; the ominous task of building up the civic realm, bureaucratic rationality, technocratic efficiency and an enabling environment still lie ahead.

Whereas the "new political sociology" firmly advocates the separation of society and social ties from economic life, many critics question the inevitability of such models of modernization, see (Amin, 1987; Shivji, 1991). The implementation of liberal democracy, multi-party politics, elections, restoration of human rights and individual freedoms while necessary and commendable are no guarantee of economic growth and development (Nyang'oro, 1994: Robinson, 1994).

Nabudere (1989) and Post (1991) argue that the liberal democratic model and forms of civil society associated with it are based on Western ideologies and are inappropriate for Africa. However, as Sandbrook (1992) and others point out; what are the alternatives given the bankruptcy and collapse of "client capitalist" or "African socialist" one party states or "statism" in its many forms?

The Gramscian notion of civil society is more ideological and less structural. Civil society often reinforces the ideology and organization of the dominant capitalist classes. Hegemonic structures in civil society reinforce and reproduce the values and norms of market economies as the common ethical values.

> In this multiplicity of private associations or civil society . . . one or more predominate relatively or absolutely . . .constituting the hegemonic apparatus of one social group over the rest of the population (or civil society). (Gramsci, 1971, p.263)

Tanzania lacks cohesive capitalist economic structures as well as an elite that consistently promotes capitalist ideology; i.e. that the class structures associated with and elaborated by the competitive capitalist markets are the best means to achieve both political maturity and economic development. Young (1994) and Callaghy (1994) question whether privatization and reform will have the unintended consequence of intensifying rent-seeking behaviour. Civil society in the Gramscian formulation can play a negative role if it supports the corrupt "predatory bloc" and masks classes and hierarchical and exploitative structures and relations.

Critics such as Lemarchand (1991) or Eyoh (1996) debunk the "convivialist fallacy", the mistaken assumption that the informal economy is not tied to the state and the predatory classes or that most economic actors outside the predatory bloc are of homogenous socio-economic status. They argue that this fallacy masks or fails to recognize new modes of exploitation, class formation and miscarriages of social justice. As Bratton (1989) points out there is a dialectic of engagement between civil society and the state.

The Gramscian view also includes the possibility of creating new social systems in which there will be support for the popular bloc and the creation and validation of "virtues of collective life" Noumoff (2000). As Bratton (1989) points out there is a dialectic of engagement between civil society and the state. Hegemony is a combat zone, African resistance movements have frequently been defeated and repressed by the predatory state (Bratton, 1994; Woods, 1992; Chazan, 1993). Bratton (1994) highlights how the patron-client ideology and practice is embedded in civil society in Zimbabwe. Barkan (1994) outlines a similar process of how ethnically divided political associations promulgate ethnic conflicts in Kenya. Unfortunately one could choose many more examples.

Pluralism requires civic action that resists authoritarianism. Many analysts believe civil society must challenge states that lack legitimacy (Chege, 1994; Chazan, 1992; Shoemaker, 1995; Chege, 1994; Sandbrook, 1992). Holm, Molutsi and Somolekae (1996) operationalize civil society in Botswana in a very similar way, they define it as; associations demanding government responsiveness to member concerns and involved in changing government policies to protect individual rights and freedoms. Shoemaker (1995, p.7) uses a by now familiar checks and balances logic in his attempt at a comprehensive definition of civil society as:

> A broad sphere of social interaction between the household and the state dominated by voluntary associations . . . that define their purpose primarily as civic action to influence state institutions and policy processes and to foster a democratic culture.

Tripp (1994) notes that the struggle to create pluralized associations in Tanzania and Uganda, and within associations, to create new rules and operational procedures, will be a painful learning experience. She applauds women's associations that are less frequently based on ethnicity and religion than most other associational life in Tanzania and Uganda. Tripp's operational definition includes informal unregistered groups that are outside the state but play a key role in reclaiming political space and providing a base for institutional change, see also Barker (1994). By including informal groups civil society takes on a "now you see it now you don't" character.

One obvious problem with the definitions of civil society that include unregistered and informal groups is that they seldom "define a purpose" in some permanent or organized fashion as that purpose can change with circumstances of the day. Perhaps more importantly informal groups may "influence the state" by promoting civil unrest, or working *against* polyarchy, accountability and constitutional means. However, since few rural Africans belong to registered associations and the majority of Africans still live in rural areas, to exclude informal groups is to severely emasculate the "master concept".

Rustow (1970) highlights another important contradiction; the factors that bring democratic reforms into existence may not be the same ones that sustain a long-term democratization process. For example, Huntington (1991) and Diamond, Linz and Lipset (1988) stress the importance of what Robinson (1994) refers to as the "conjuncture". Robinson's excellent discussion of National Conferences in Francophone Africa, notes wide ranging external influence from neighbouring countries, from South Africa, from France and French historical conditions, from Russia and Eastern Europe and so on. Such diffusion of democratizing influences is an important part of *regime change* but the dilemma of sustainability of democratization initiatives is still very much an open question.

As Chatterjee (1990) points out most of the newer literature on civil society tends toward ethnocentrism since a central assumption is that all societies must replicate European types of polyarchy, transparency, accountability and constitutionalism. This is the form of Afro-pessimism that Eyoh (1996) labels teleological.

Not only the predatory state, and rent-seeking elites fail to measure up but so does civil society. Tripp (1994) gives an excellent example of how the norms of accountability and civic responsibility were violated in an association supposed to represent disabled women in Uganda. As one member stated "Our leader was unfair and gained a lot from us . . . she offered herself the aid meant for us".

Gramsci demarcated "civil society" from "political society". The latter comprises legislation and implementation of legislation, political parties, elections, the armed forces, the police and other agents of state repression. Civil society reinforces the ideological and organizational structures of the dominant capitalist classes in capitalist societies.

Civil society under capitalist hegemony experts pressure to preserve the prevailing ideologies and forms of consciousness. Examples abound in Africa: Bayart (1992) demonstrates how the popular classes often prefer the structures of domination to the unknown realms of freedom.

Lemarchand (1995) and Newbury (1995) highlight how some forms of popular action in Rwanda and Burundi degenerated into incivism, ethnic conflict, crimes against humanity and genocide. Roitman (1990) and Ninsin (1988) note how cycles of recrimination and violence perpetrated by the state tend to push civil society actors to respond in kind leading to further waves of violence. (Stamp, 1991; Meeker and Meekers, 1997) portray how gender discrimination, harassment of women, and patriarchal norms and customs operate both in the state and civil society. Women politicians and activists in associations are often subjected to harassment and violence.

The Gramscian critique is not entirely pessimistic since the present civil society and political society may be replaced by more humanistic development. Gramsci highlighted the need for cultural change associated with liberal democratic reforms and freedoms.

> . . . an element of active culture . . . a movement to create a new type of man and citizen . . . to determine the will to construct within the husk of political society a complex and well articulated civil society in which the individual can govern himself. (Gramsci, 1971, p.268) . . . It is possible to imagine the coercive element of the state withering away by degrees as ever more conspicuous elements of regulated society (or the ethical state or civil society) make their appearance. (Gramsci, 1971, p.263)

However will the "ethical" elements within African civil society be capable of promoting appropriate African systems of governance based on social justice?

African cultures legitimate local knowledge and can validate appropriate political practice. Tripp (1994) gives the example of a Ugandan women's group whose name when translated into English means "do not dismiss what appears to be insignificant". Even when the popular classes appear dormant and without a voice, resistance theory looks into actions and communications, contested meanings, habits, customs and symbols that legitimate or undermine the power structures

(Boulaga, 1993; Schatzberg, 1993) practice that can transform structures (Bourdieu, 1977).

Almond and Verba (1965) discuss political culture as a structure that maintains social equilibrium, whereas resistance theorists regard culture as a force that makes societies and social structures dynamic. The "culture of politics" (Robinson, 1994) or "cultural schemas" (Ortner, 1989) or "cultural scripts" (Geertz, 1990) operate to transform societies in certain *conjunctures* or circumstances: "the notion of structure as constraint is turned on its head" (Robinson, 1994) as cultures and social systems transform themselves under both internal and external prompting. Robinson (1994) delineates how the National Conferences in Francophone Africa demonstrated "discontinuities of values" (Bermeo, 1992) leading to social change.

In the conjuncture of the early 1990s, cultural schemas from differing geographic regions became "strategy generating principles" as ideas and examples of social change and support poured in from France, South Africa, Eastern Europe and elsewhere and were taken up by African leaders and political associations to lobby for democratic reforms. As Ake (1992) and others note there are deep African cultural roots for broad political participation and political accountability.

Resistance theorists all agree that civil society plays a critical role during periods of democratic openings (O'Donnell and Schmitter, 1986; Bratton and Van Der Walle, 1994). Different labels are used to refer to such openings: "political openings" or "political opportunity structures" or "power asymmetries" (Robinson, 1994) or "political space" (Barker, 1994). They usually demarcate civil society as groups of people and institutions that are not part of the state (Chazan and Rothschild, 1993; Nyang'oro, 1994). In this broad sense civil society becomes "potentially subversive" political space that may topple authoritarian/ predatory regimes (Fatton, 1992).

When groups of twenty or so villagers meet in the *ekigata,* the local informal sector bars in the Great Lakes region, and the patrons heatedly discuss politics: this may seem insignificant, but such micro-politics are the most common the way villagers meet and exchange information (as discussed in another chapter). Villagers obtain international information from those who listen to short wave radio and sometimes get to meet relatives home from the larger cities; they have a chance to debate and express opinions in a highly interactive format, form opinions and then use these beliefs as a guide to action.

As Aronowitz and Giroux (1993) point out domination is never static or complete, resistance movements develop within emancipatory and oppositional public spheres. It is sometimes possible to replace alienation and alienating need structures with liberatory practices; however, resistance theory also acknowledges that not all oppositional behaviour and resistance is progressive. The cultural schema of the oppressed may also reproduce structures of domination such as patriarchy (Sisulu, et al., 1991). Civil society contains many internal contradictions and "marginal" as well as "modal" practices (Robinson, 1994).

Civil society may help to undo the detrimental effects of state-centred accumulation, however there are contradictions within and among classes outside of the predatory bloc. As Fatton (1995) points out the three primary sectors in African political economy (the predatory elites, the middle classes and the impoverished urban and rural popular classes) are all mirrored in civil society. Different formal and informal associations represent different blocs. Movements promoting social justice and the eradication of poverty face the double challenge of escaping from chronic debt and poor economic performance and resolving existing and emerging class contradictions. Also, as popular forces in civil society resist the hegemonic ambitions of rent-seeking politically connected elites, they are perceived as dangerous and subjected to repression and cooptation by the predatory bloc (Sandbrook, 1988; Barker, 1994).

Despite the fears of Eyoh (1996) and others that the struggle for liberal democracy is essentially a middle class phenomenon and does not represent the impoverished majority, it may be the only alternative and therefore the only way to move forward (Ibrahim, 1986). As associations gain autonomy and win the struggle for individual freedoms this creates possibilities for widening the struggle and for the oppressed to gain a voice and carve out their own political space. Middle class leadership may also help separate the economy from the ravages of rent-seeking, politically-connected, predatory elites and the corrupt patronage of "state centred accumulation".

"Contested meanings and power asymmetries" will not inevitably lead to social justice: however resistance theory at least allows for hope and the "language of possibility".

Classes and Hegemony

What is the connection between liberalization, classes and class formation and civil society in the Great Lakes region of Tanzania? It is too early to make any conclusive predictions however we can at least highlight the changes in the structure of the elites within the country and the region. The three social groups who dominate the economy and society must be transformed if liberalization is to succeed in a sustainable way. These groups are not clearly defined by ownership of the means of production or other classic demarcations of class.

The first group is the aid establishment. In the 1970s and 1980s Tanzania was one of the world's most aid dependent countries; since about two-thirds of national wealth came from foreign loans and grants. Foreign nationals who controlled development assistance, exerted a very strong influence over each project and ultimately over the commanding heights of the economy. Unfortunately there was little coordination between projects and few projects were sustainable without further aid: the end result was the economic crisis of the 1980s.

The second group are the Tanzanian nationals who control the state: in Tanzania this included parastatals (such as some banks, most official transport, postal services, electric power, marketing boards, and a wide range of enterprises).

Rich farmers are the third and least visible elite: they operate most of the informal sector transport, trade, and agricultural production. The rich farmer class produces goods and services that are not dependent on foreign assistance.

The Commanders of the Commanding Heights

Bretton Woods is a scenic resort in the shadow of Mount Washington in New Hampshire's White Mountains. In July 1944, US President Roosevelt and other world leaders met there to set up the monetary system of the post war world, creating the leading international financial institutions, the World Bank and its sister institution the International Monetary Fund. For the first five years the Bank and the Fund focussed their efforts on the reconstruction of war torn Europe, but by 1949 the World Bank's second President John McCloy remarked "The reconstruction phase of the Bank's activity is largely over and the development phase is underway". Yet over forty years later, Hancock's remarks (1991, p.190) could justifiably apply to Tanzania;

> ... the only measurable impact of all these decades of development has been to turn tenacious survivors into helpless dependents [and p.189] ... the obvious conclusion of this book: that aid is a waste of time and money, that its results are fundamentally bad, and that — far from being increased — it should be stopped forthwith, before more damage is done.

Although Hancock's intention was to shock the public and perhaps shake-up the aid establishment the sentiments he expresses did become a part of the public domain and the publication of *Sub-Saharan Africa: From Crisis to Sustainable Growth* by the World Bank in 1989, signals a change in philosophy within the Bretton Woods institutions and other powerful international financial institutions. Sandbrook (1995) uses the term "structuralist radicals" to label the pro-liberalization faction, ie. those decision-makers who believe that economic development requires transparency and accountability and this will only be possible *after* significant political reform and democratization (Sandbrook, 1995).

In an earlier work Sandbrook (1992) questioned the effects of aid in Africa where: agricultural output has stagnated or declined, industrial output has fallen, desertification and deforestation are common, food imports are rising, terms of trade are declining, more capital leaves the continent than comes in, infant mortality rates are on the rise, nutrition continues to decline and the middle classes are becoming pauperized. By 1990, servicing Africa's external debt required half of all export earnings. Hancock's "helpless dependent" scenario becomes the daily reality for millions of Africans, as Sandbrook (1992, p.5) avows

> ... the number of Africans, enduring absolute poverty grew by almost two-thirds in the first half of the 1980s ... sub-Saharan Africa will suffer an increase of 85 million in the numbers of the poor by the year 2000.

For Hancock (1991, p.192) the primary villains are "that notorious club of parasites and hangers-on made up of the United Nations, the World Bank and the bilateral agencies". The system bypasses the poor: Hancock (1991, p.190) probably consciously overstates his case:

> ... most poor people in most poor countries most of the time *never* receive or even make contact with aid in any tangible shape or form: ... After the multi-billion-dollar financial flows involved have been shaken through the sieve of over-priced and irrelevant goods that must be bought in the donor countries, filtered again in the deep pockets of hundreds of thousands of foreign experts and aid agency staff, skimmed off by dishonest commission agents, and stolen by corrupt Ministers and Presidents, there is really very little left to go around. This little, furthermore, is then used thoughtlessly or maliciously, or irresponsibly by those in power — who have no mandate from the poor, who do not consult with them and are utterly indifferent to their fate.

Since the aid agencies are not adequately accountable to taxpayers in developed countries they end up with poorly designed policies — kowtowing to the ascendant political climate in the donor countries whether "welfare statism" or "resurgent conservative values" or the creation of an "enabling environment" through economic and political reform conditionality. Hancock (1991, p.73) lists a smorgasbord of inconsistent and contradictory aid industry beliefs: that includes "trickle down" and "bottom up" initiatives, that support hi-tech industrialization, that then switch to agriculture as the economic base of the majority, that claim the best way to boost agriculture is to support large-scale farms, but at other times claim that agricultural progress requires assisting small farmers, that promote state control and central planning and then turn this around with policies that support entrepreneurs who should be given a free hand in a free enterprise system. Tanzania, in almost four decades of independence experimented with all the above policy options and it is hardly coincidental that Tanzania has been one of world's most aid dependent nations, which in 1988, derived about 60 percent of its Gross Domestic Product from foreign loans and grants. Even in 1998, Tanzania's external debt stands at eight billion US dollars and some analysts consider the aid donors as Tanzania's hidden ruling class though the vast majority of them are only temporary residents or visitors. Over thirty bilateral and multilateral aid donors (such as the IMF and the World Bank) provide this official development assistance. The agents who control most of these funds follow their own agenda and are effectively cut off from the Tanzanian populace by barriers of nationality, language, culture, lifestyle, wealth and often race.

Bilateral agencies usually promote policies and development strategies that serve their own national interests or at least focus on national specialties. As Julius Nyerere pointed out in his famous pamphlet *The Arusha Declaration Ten Years Later,* Tanzania lacked a coherent planning policy and accepted many projects that rarely fit together very well. The national railway system, for example, uses two different gauges of track and many complications arise because rolling stock is not compatible. Besides a lack of holistic development planning, few projects were designed to ensure that they would be able to pay their recurrent expenses once the aid donors withdrew, and so many fall into disrepair once the donors leave.

Multilateral assistance, is usually tied to policy guidelines such as the recent IMF-imposed structural adjustment program. In theory these programs are designed to trim the costly and ineffective public sector and allow local and foreign entrepreneurs a freer hand in restructuring the economy and making it sustainable. This process may be facilitated by recent discoveries of mineral wealth. Multinational mineral exploration companies have agreed to spend millions of dollars to develop off-shore oil and natural gas reserves and to set up large scale gold mining near Lake Victoria. Yet there are two nagging problems that the SAPs may not solve. The first is how to improve the infrastructure so that investment becomes more attractive, if this requires more development loans then the debt burden will become even heavier. The second major problem is that the reduction and elimination of social services may incite tremendous opposition. The SAPs have drawn critical fire from local and international activists because they seek to reduce deficits by cutting social spending, using cost sharing programs, and promoting increased production of cash crops for export. These policies are at odds with the needs of poorer and middle status farmers who no longer receive affordable medical and social services. Poor villagers cannot afford the medical consultation and other fees resulting from cost sharing and cut-backs. Retrenchment often resulted in state-run hospitals and clinics with virtually no drugs or doctors and abysmal standards of sanitation and lower school enrolments because parents cannot afford to apy school fees.

The SAP's reliance on export crops has obvious pitfalls. Tanzania's largest export is coffee: in periods of slumping world coffee prices it does not compete effectively with informal sector activities.

Another persistent problem is what Sandbrook (1992) calls the fiscal crisis. Since the national economy relies very strongly on taxes derived from exports, if peasants shift their resources from export crops to localized markets then the government will not collect enough revenue to run its development programs creating a vicious circle of discontent.

In Tanzania where almost nine out of ten citizens are small farmers, unpopular agricultural policies have generated popular resentment that led to the first multiparty election. The Rwanda crisis and the influx of over one-half million refugees into the region created additional mega problems. Obviously very complicated power struggles were going on between the UNHCR led aid establishment running the refugee camps and disbursing millions of dollars, and the national government. The fact that the Rwandans were expelled from Tanzania indicates that the agents of development assistance are limited in the elite role of influencing state policy and that the national political class are not always figure heads or puppets the dependency model usually implies. However, environmental damage caused by the refugee crisis and the flooding from the heavy El Nino rains in 1997 and 1998 demonstrated clearly that the national government is dependent on international agencies for disaster relief during times of crisis.

The issues are obviously complex, one must be careful in apportioning too much blame to either the aid donors who do often foster dependency or the "neo-patrimonial" state who do often waste resources through corruption and poor governance. Whether or not a class of entrepreneurs and bougeois industrialists will emerge is an open question that depends on the trajectory of economic development and the creation of a much larger national market. Neither foreign nor domestic capital can at present mobilize the development resources that would create an industrialized nation. Even Tanzania's more developed neighbour Kenya is unlikely to reach its objective of industrialization by the year 2020.

The Tanzanian National Elite

The national elite or governing class came to power at independence and under Julius Nyerere did generate a good deal of popular enthusiasm, support and legitimacy. However as Hyden (1994) correctly points out the economy stagnated under the "one-party socialist" state and Nyerere used his popularity and legitimacy to either coopt civil society into the governing class or completely marginalize it. Thus the governing class while *politically hegemonic* was caught in the contradiction of being the agents of unsuccessful economic policies often caught in the role of counterparts or simply underlings to the agents of development assistance (Von Freyhold, 1977, p.85). The national elites control the post colonial state but govern under the scrutiny of international agencies. This class can coerce the peasantry but it cannot take charge of actual production or even tax peasants effectively. Bernstein (1981, p.54) argued that the state, had "no base in civil society" and therefore created "its own social base through the sheer expansion of the numbers of people it employs". Although Bernstein writes from a Marxist perspective, his conceptualization of the separation of state and civil society is very similar to a variant of modernization theory, Hyden's notion of the African state "suspended in mid-air".

President Nyerere virtually eliminated civil society and parliamentary opposition to the party with the Arusha Declaration in 1967 and the constitutional reforms of 1977: however in the early 1990s Nyerere was the first prominent political leader to raise serious doubts about the one party system. In 1993 President Mwinyi appointed a constitutional review commission under the leadership of Chief Justice Francis Nyalali and its report recommended the introduction of multiparty elections that was soon enacted into law and implemented (Hyden, 1994). Without overt external pressure such as withholding assistance (as was the case in 1992 in Kenya) the Tanzanian state transformed itself and helped to reestablish civil society. Essentially the governing class had little choice since it was an artificial creation to start with. It had expanded employment in both the civil service and parastatal sectors creating institutions that were inefficient to the point of reducing production to unacceptably low levels and creating a very high degree of excess capacity. Ultimately, the political fraction, including unproductive parastatal enterprises, is dependent either on foreign aid, which can be capricious or else they must squeeze smallholder agriculture to provide revenue. In attempting to augment peasant surplus production, this group alienated peasants, and the middle classes, caught in a muddle of red tape and restrictions. Predictably the result was growing debt and crisis as production and state revenues were reduced during the economic stagnation of the 1980s.

Prior to the constitutional amendments in 1992 allowing multi-party elections, Tanzania could be characterized as "state capitalism" (Von Freyhold, 1981) or "one--party socialism" (Barkan, 1994). However regardless of the constitutional system, the social and economic interests of the peasantry tend to be structurally opposed to those of both the governing class that controls the state and related enterprise and the aid providers. Tanzania is an agricultural country, the entire urban sector is only about ten per cent of the population, and therefore is too small to sustain itself. Bernstein (1977 & 1981) refers to the "simple reproduction squeeze": smallholder households are being pressed to provide surplus (as taxpayers and as underpaid producers) to the point that even their basic subsistence (simple reproduction) is being threatened.

Deborah Fahy Bryceson (1990) supplements the above analysis by arguing that the Tanzanian state (like the chiefs of an earlier period) sustained their legitimacy through "parastatal clientage" — a term she uses to describe particularism and corruption throughout the state run sector. She argues that the state apparatus of the 1980s was cut off from local rural clientage networks and was no longer based on tribalism, instead the urban Dar es Salaam elites, manipulated the state bureaucracy to benefit themselves, their client employees and the extended family networks. Munishi (1989) calls the Tanzanian political system "bureaucratic feudalism", Sandbrook (1992) and others refer generally to African political cultures as "neo-patrimonialism", Fatton (1994) talks of the "predatory bloc" or "pre-bendalism", Bayart (1993) discusses the "politics of the belly", Eyoh (1996) introduces new terminology "a rhyzomic system of patronage" etc. but whatever terminology is used, the net result is the same. Sandbrook (1993) describes the collapse of the civil service and the breakdown of professionalism in many African countries. This is combined with the disintegration and near breakdown of official economies as parastatals and public services performed extremely poorly although often overstaffed. One can also perceive an underlying tension between organized ruling political parties, the press, political opposition parties and nongovernmental organizations (Tripp, 1994).

Elsewhere in this book I note that none of the resident family members of the two hundred and fifty household sample reported being active members of a political party or of a voluntary organization that was active in reform politics. Male heads of household met frequently at informal sector bars and discussed a range of topics including politics. Such discussions would of course influence voting choices and other political behaviour. However, there is no evidence to support the civil society model that presupposes that all strata of society participate in voluntary associations that discipline the state. In this formal sense rural people in the Kagera region are largely disenfranchised; nevertheless if civil society includes informal association then an important process of creating "civism" or the "civic realm" is at work as village people meet frequently and socialize with their neighbours and exchange ideas that diffuse into the rural scene from the national and international arenas.

When citizens who live in rural areas exercise their voting rights or in infrequent cases engage in direct action, then even isolated villagers have an impact on the state and its performance and will help to resolve the key questions. Can the new supposedly reform-minded state reduce its size while at the same time improving the quality of public services? Can it curb predatory excesses, and transcend the embedded systems of patronage that also prevails in registered voluntary associations cf. (Bratton, 1994; Tripp, 1994; Fatton, 1995). How can a formidable opposition movement emerge let alone succeed? Can it successfully challenge structures that are embedded throughout the state and society, that Eyoh (1996, p.61) refers to as "a generalized contempt for public law"? Can the reform/ democratization movement somehow collude with the private sector to reduce clientage and the rent-seeking investments of African elites? These structural dilemmas seem to be the reason that well-respected political analysts such as Ali Mazrui in his address to the OAU and William Pfaff in *Commentary* and other venues suggest, perhaps in a mood of despair, recolonization for some of Africa's basket cases.

The Peasant Elite

The richer farmers oppose the state and its particularistic clientage networks for three reasons. First, all strata of the peasantry, especially the rich, want to obtain higher producer prices for officially marketed crops, putting them in direct competition with the urban political elites. Second, the political may compete with rich farmers as a rival for leadership over the peasantry. Bernstein (1981) demarcates the rich farmer power base in the cooperative societies and the church, whereas Bryceson (1991) argues that village and clan ties define the clientage base of the rich farmer class. Third, they earn relatively large incomes in the informal sector providing transport and consumer goods often in competition with official initiatives and legally registered businesses.

In the world of abstract class analysis there is a split between the two national elites who are clearly separate structural entities. In the real world, this split may hardly be perceptible. As Fatton (1995) points out it is often wealthy politicians or their associates, members of the "predatory bloc" who control or at least derive the largest share of profits from the informal economy including rural enterprise. Reintegration into rural society may occur at different stages in the careers of the politically connected, since the "crony class" usually retire to become rich farmers, however it may occur earlier as agents in the state and the official economy also dabble in the informal sectors.

Even if the roles are contradictory, individual urban migrants must plan for their retirement (officially at age fifty) and look after their future interests. (In later chapters I demonstrate how the urban migrant group are the largest occupational category to become *rich* farmers). One head of household in this sample had been an important government official before retiring as a rich farmer. Moreover, in later chapters we see clear evidence of class formation within the peasantry and this of course raises the possibility of new forms of exploitation that belie the assumption of social homogeneity.

One thing urban elite and rich farmers have in common is their high socio-economic status that separates them from the masses of the peasantry with whom they compete for resources. Von Freyhold gives examples of richer villagers monopolizing resources and also "hijacking" communal projects such as bars or shops, which they take over when they become successful. In other cases communal enterprises were forced out of business by competing privately owned operations (Von Freyhold, 1981, p.88).

The political fraction is alienated from peasants because they collect taxes and set producer prices at a level below that which most farmers consider fair. Also, most peasants believe that the taxes that they pay are wasted, embezzled or at best spent on projects that benefit other classes. In our study we asked small farmers if they thought the development tax was well spent; only two percent said they thought it was. From the point of view of the two hundred and fifty households in this study the national and international elites and the programmes that they generated were marginal to their day to day survival: they received little if any benefit from either.

Conclusion

Are the above conceptualizations of political culture useful or do they merely add to the confusion? I hope that they can be viewed as mutually reinforcing. Clearly there is much merit to the Weberian notion of building a civic realm where bureaucratic rationality and transparency and accountability prevail. Tanzania is rife with reports of corruption: For example corruption at the port of Dar es Salaam, corruption in the legal system, the Warioba report on corruption implicating many high government officials, the visit by Robert MacNamara's corruption watchdog, the Global Coalition for Africa to investigate tendering irregularities on government contracts, and the response by the President himself, forming and chairing an investigative commission on tendering procedures, cf. *The East African,* Nairobi, October 20-26, 1997, January 19-25, 1998, March 2-8, 1998, March 9-15, 1998, and so on. Poor governance, neo-patrimonialism and clientage networks are major structural problems that must be overcome if reforms are to work.

As Von Freyhold and others demonstrate Tanzania with an external debt of eight billion US dollars at the beginning of 1998, is very susceptible to pressure from the IFIs and other donors. In the present conjuncture some civil associations and most donors share a vested interest in making the system less corrupt and more workable. The success or failure of liberalization and the drive for transparency and accountability provide a litmus test of the power of the agents of "development assistance" and whether a discursive network of expatriate technocrats can be replaced by nationally based entrepreneurs.

The grass-roots village-based innovations presented in later chapters demonstrate the key element of resistance theory "do not dismiss what appears to be insignificant". It is not only the advocates of the theory of the "uncaptured peasantry" who believe peasants are backward and non-scientific, many well-to-do farmers expressed the same opinion about their village neighbours. However this opinion may merely serve to mask class interests. Is it not more reasonable to assume that "the farmer ... knows his business better than we do, unless there is evidence to the contrary" (Hill, 1986, p.28) and, empowerment is what peasants really need? Is it not the bureaucrat who must learn how to incorporate the farmer into the process of innovation and development by offering *appropriate* assistance? As we will see, some farmers who can obtain even a very small amount of capital, at the right time in their career cycle, can go on to become prosperous.

Tanzania could give higher priority to providing logistical and financial support to innovations already used by smallholders. International aid-givers, politicians and bureaucrats, the richer farmers, and the peasant hearth holds (Guyer, 1986) all share a common interest in sustainable development. The most difficult task will be for these groups to reach compromises on issues where there is a conflict of particular interests.

Conflicts between small farmers and the state over the division of export crop revenues, may be amenable to compromise solutions. If the bureaucracy sets higher producer prices for coffee they may optimize revenues by increasing supply. The government might also benefit by providing increased support to progressive individuals and associations because these are usually more cost effective than larger estates especially parastatals.

Can national organizations, and the new wave of corporate investors cooperate with localized farmer initiated innovations in transport and trade? Can such projects raise real incomes by providing not only cash but consumer goods on a regular basis and at low cost? Ultimately if the Tanzanian state is to overcome its continuing "fiscal crisis" and succeed in its proposed tax reforms, it must encourage the peasants to increase their involvement in the official economy and provide material incentives for them to do so. But the agents of the IFIs and national government may need a push from the popular bloc of civil society that will provide the social force, the political will and the economic means so that the system does become more accountable and an "enabling environment" is created and the nation can prosper and resolve its enormous economic problems.

4 The Ecology of the Kagera Region: The Physical and Human Environment

Introduction

The Haya farming system has proved to be sustainable for at least four centuries because it was developed in harmony with local resources and ecology. Rural population densities are much higher than in most other areas of Tanzania, where survival requires shifting cultivation, long fallow periods and a widely dispersed population (Bryceson, 1990, p.23). Haya residents live on permanent plots where they cultivate bananas as the staple food. This is possible because Kagera region is endowed with ample rainfall and deep soils, moreover, the Haya have a long tradition of farming in a manner that preserves and enhances the fertility and structure of the land as it was in its forest state prior to settlement. This is achieved through interplanting, manuring and mulching.

The Haya farming system has advantages frequently absent in African agriculture: reliable rainfall, a reasonably good system of environmental preservation, a tradition of modern education stemming from an early missionary presence. Unlike Kagera Region's neighbours, Uganda, Burundi and Rwanda, Tanzania is a stable and peaceful nation that has avoided the traumas of civil war and ethnic clashes. However Kagera also has certain disadvantages: it is cut off by 1500 kilometres of bad roads from Dar es Salaam, the main port and the national commercial and political centre; its intra and inter-regional transport and communications infrastructure is very poor; and it tends to be ignored by the state and left out of national development initiatives, perhaps because of an ongoing tradition of dissent.

The Haya live on the Eastern frontier of the "interlacustrine" or Great Lakes region of Central Africa; an area approximately 350 kilometres wide and 700 kilometres long. The zone is bounded by Lake Albert in the Northwest, Lake Tanganyika in the Southwest and Lake Victoria in the East. The major ethnic groupings in the region are Nyoro, Ganda, Soga, Toro, Ankole, Haya/Nyambo, Hutu, Tutsi, Zinza and Ha (Reining, 1967, pp.15-15a).

The Great Lakes area has been the site of "significant vegetational change" for the past 3000 years when "forest decreased and grass increased" (Schmidt, 1979, p.24). This change is the result of agricultural activity in which the natural fertility of part of the land (used for permanent banana plantations) was maintained and often enhanced; while the land devoted to pastures and seasonal crops has been degraded, with resulting loss of fertility and patches of laterization (laterite clay dries into a brick-like surface).

Kagera region measures 28,750 square kilometres, the altitude is 1100 metres at the Lake Victoria shore rising to 1600 metres along ridges which form a step-like pattern, extending inland, up the West side of the Great Rift Valley.

Further West is the drier Karagwe plateau which extends up to the base of another ridge system beginning at the Rwanda and Burundi boundaries. Each system contains large escarpments, plateaux, rivers, lakes and swamps (Rald and Rald, 1975, p.30; Boesen et al., 1977, p.22).

Conyers (1971) identifies twenty micro environments in the Kagera region, but many of these are swamps and forest reserves; marginal areas with few permanent residents. However, there are five *major* groupings:

1) In the North, bordering Uganda and Rwanda, the Kagera river has carved out an area of steep canyons and ravines: this is neither a major centre of population or agriculture, although there is one large (8000 hectare) project the Kagera sugar estate. The Kagera and its tributaries provide a natural border with Rwanda and it became notorious during the genocide when thousands of bodies, found floating down to Lake Victoria were removed and unknown others reached their grisly destination.

2) In the Southeast between Biharamulo and Mwanza, unoccupied farmland is rapidly being settled, although because of a

poorly developed infrastructure, it is still a peripheral part of the cotton growing zone centred in Mwanza. However, it may soon become a part of a gold mining boom, in early 1996 Sutton Resources of Canada invested $187 million to carry out further gold exploration and to develop its Forest Reef Mine on the Ikungu Peninsula, *The East African,* Nairobi, June 17-25,1996.

3) The Ngara district in the Southwest is ecologically similar to the central banana/coffee zone, but it is isolated by four hundred kilometres of bad roads from Bukoba or Mwanza. Ngara is more easily accessible to Rwanda and between April 1994 and the end of 1996 there were four large refugee camps where about one-half million Rwandan Hutus fled. The largest camp, Benaco, was a tent city of over two hundred thousand mostly Hutu refugees. This massive relief operation coordinated by the UNHCR, United Nations High Commissioner for Refugees, brought together a host of relief agencies and other NGOs, nongovernmental organizations agencies, such as OXFAM or CARE. Whether the millions of dollars pumped into this region to care for refugees will offset the destruction of millions of trees and the using up and pollution of water resources remains to be seen. The Rwandan debacle may permanently alter the regional *ecology and economy* of this area.

4) Karagwe is a plateau formation west of the coastal ridges where, in days of past glory, the population densities were high, huge herds of cattle grazed on the grassland and its kings were in control of large armies. The herds were decimated by the rinderpest epidemics which accompanied German conquest in the 1890s and the area became poor as large numbers of its people migrated. It is drier than the coastal ridge zone and relatively underpopulated. There were an additional 100,000 Rwandan refugees in Karagwe camps that straddled the western boundary between Tanzania, Rwanda and Burundi. The refugee camps in the Ngara and Karagwe districts provided many marketing opportunities for food and cash crops (other than coffee) but how much long-term damage was caused by these settlements?

5) The coastal ridge area is the economic centre of Kagera region:

it contains three administrative units, Bukoba Urban, Bukoba Rural, and Muleba districts. About sixty percent of Kagera's population live in the coastal area, and this ecological zone produces sixty-two per cent of the West Lake area's total agricultural production and eighty-five per cent of the region's coffee — the major source of foreign exchange (Tibaijuka, 1979, pp.32, 44).

The coastal ridge is the regional centre of agriculture, trade, fishing, construction, transport, and craft industries. Outside the coastal zone, twelve per cent of the total land is cultivated while inside this zone, permanent cultivation covers about one-half of the land, although micro studies note slight variations, for example: thirty-nine per cent of the land occupied by banana groves, and another eight percent comprised of pasture and annual crop areas (Kabwato, 1976, p.84); forty-eight per cent of the land is cropped (Tibaijuka, 1979, p.20); forty five per cent under cultivation (Rald and Rald, 1975, p.19); forty-nine per cent cultivated (Rald, 1976, p.25). Land-use surveys clearly indicate that under the present system all the land which can sustain agriculture is being farmed. Other land can only be used for grazing or as a reserve.

The farmers of the coastal ridge system divide their land into four types: 1) permanent groves of bananas interplanted with coffee and beans, 2) small open patches of annual crops such as yams, maize and groundnuts — planted in mounds which are allowed long fallow periods 3) communal cattle pastures 4) communal bush land used for firewood and swamp grass. Agricultural labour on the four types of land is seasonal, it will be discussed below under the rubric of the "ag ricultural seasons" and cattle tending will be discussed separately.

Kagera Climate

The regional climate is characterized by moderate temperatures and high and even rainfall. This gives Kagera's farming system a large advantage over most of Tanzania where the farming is carried out in areas which require shifting cultivation because of erratic and inadequate rainfall. Bukoba is cooler than the other Lake Victoria port cities such as

Mwanza, Kisumu or Kampala. This is because prevailing winds blow across the lake and bring more precipitation and relatively cool weather, even though the region straddles the equator, at latitude 1,00'S to 2,15'S. The Bukoba regional weather station recorded (1945-63) monthly means of 20 degrees Celsius with a daily variation between 15 and 25. Higher points are cooler with a range of plus or minus four degrees and their rainfall varies more making the ridgetops drier than the coast.

> Most of the precipitation falls as thunderstorms (mainly in the morning hours), and in this area the interrelation between atmospheric processes and the relief gives rise to very extreme local precipitation. (Rald and Rald, 1975, p.6)

Bukoba town's annual average rainfall (over 2000mm) is the highest in the region: whereas, the Igabiro farm school recorded averages of 1070mm, typical of the higher coastal ridges. (In Tanzania fifty-five percent of the country receives less than 750mm per year). Kagera's rainfall is adequate throughout the year, and is high enough to support permanent cultivation of plantain as the staple food. The dry season lasts from June to September with July as the only really dry month. The rest of the year is wet with two peaks, a major one between March and May and a minor one between November and December (Rald and Rald, 1975, pp.11-17, provide detailed climate charts). There are annual variations that threaten the banana crop, about once in a decade. Kagera's worst climatic disaster this century was caused by the "El Nino" rains in late 1997 and early 1998; these caused massive flooding throughout East Africa. Bukoba District was one of the worst hit: floods left 13,000 people homeless and destroyed 8000 hectares of banana/ coffee plantation, *The East African,* Nairobi, January 19-25, 1998.

The Land

Typical of the regional topography are the wide, flat and relatively fertile sandstone ridges (separated by cliffs) running North to South, parallel to

Lake Victoria.

Regional soils can be very poor and nutrient deficient (Reining, 1962, p.60). Mbilinyi (1968, p.128) reports that the heavy rains cause leaching, poor cation exchange, a shortfall in sodium, potassium and magnesium, and high acidity levels with ph. values averaging 4-4.5. Tibaijuka (1984, p.51) cites poor availability of phosphorus to plant roots even though it is present in sufficient quantities. Soils derived from parent materials of coarser quartz and silica sandstone are much worse than those from parent materials of superior shales and finer sandstone. Milne (1938, pp.16-17) and McMaster (1960, p.85) estimate that the better soils cover one-fifth of the region. The best soils are deep (usually about two metres) loamy sands or sandy clay ferralsols located on the plateau tops (Rald and Rald, 1975, p.11; Tibaijuka, 1979, p.15).

Over centuries, the people of the region learned to identify fertile forest areas with reliable rainfall, and it was in such spots that they established permanent banana plots. The region's farmers developed ways to recycle soil nutrients and minimize leaching by planting deep rooted bananas and trees, such as, *Markhamia and Maesopsis Eminii* (Milne, 1938, p.19). Besides providing the staple food, the banana tree plays a major ecological role by shading the soil as well as helping to circulate soil nutrients. Tibaijuka (1984, p.119) lauds the banana as a soil and microclimate regulator.

Haya farmers developed additional ways of enhancing the quality of the topsoil, such as building the family house in the centre of the banana plantation, recycling household wastes (nightsoil, ashes, and kitchen refuse) and intercropping nitrogen-fixing legumes (beans) throughout the permanent crop areas. Haya farmers keep their permanent crop areas weed free and mulched with swamp grasses (when available) or banana stem cuttings and leaves. Mulching replenishes the humus layer, retards evaporation of water, prevents volatilization of soil minerals and reduces laterization.

The Agricultural Seasons

The Haya farming system has, by necessity, evolved in harmony with the land and the seasonal weather variations. The Haya divide the calendar year into four agricultural seasons: *Akanda*, mid December to mid March; *Etoigo*, mid March to mid June; *Ekyanda,* mid June to mid September; and *Omuhaguko*, mid December to mid March.

Akanda is one of the busiest seasons, the time when the Haya harvest beans, cassava, maize, and in many cases groundnuts (peanuts) and bambarra nuts. The harvest requires related activities such as drying, threshing, storage, and replanting: most of this work is done by women. The men plant new coffee seedlings and banana cuttings: they dig deep holes, pull out the old plants, shake the earth from their roots, burn the dead plants, fill up the holes with the sifted earth and ash, and if available, they shovel in composted manure or coffee hulls. If time permits they dig holes for bananas which they will plant in another season. *Akanda* is also the time when men prune the banana trees.

Etoigo, is the season of the heaviest rains, when men, especially, do the least work. Women continue with most of the weeding during this season; this is a tedious and time consuming job. Women harvest sweet potatoes in their *omusiri* or mounds of earth that are carried to marginal land: these require long fallow periods and frequent crop rotation. At this time of the year one-quarter of Haya farms plant Arabica coffee and some Robusta coffee growers use this period to prune their trees.

Ekyanda is the season of the robusta coffee harvest in July and August. It is the busiest time of the year for all members of the family and the time of peak labour demand for women. The poorest smallholders with few of their own coffee trees, may earn wages by harvesting the crop of their wealthier neighbours. This is also the harvest season for sorghum (a crop usually only grown at Lake elevation). August, is the month when the bean crop is planted in the *kibanja* (the permanent household plot) and women also cut grass from the communal pastures to be used for mulch on the core plot. *Ekyanda*, the driest and busiest time of the year, is also the peak period for beer and banana juice consumption.

Omuhaguko (September to December) is the time for planting beans, and the period of most intensive weeding. This is the time for men to prepare the deep holes for banana and coffee trees and plant more bananas. By December, beans and maize are again harvested, completing the cycle.

Cattle and the Regional Ecosystem

Cattle, an integral part of the Haya farming system, are raised on land that cannot support extensive crop cultivation. However, perennial crops require manure to maintain soil fertility and structure on the permanently cultivated plots. Farmers use manure sparingly and efficiently, they mix ash, mulch and composted manure into the holes in which coffee and banana trees are planted. Farmers who optimize their use of manure can produce yields up to five times higher than their neighbours who cannot afford cows.

This land use system evolved gradually: about five hundred years ago, there were no cattle and the less fertile areas were not farmed but preserved as game and firewood reserves (Tibaijuka, 1985). Later, as cattle were integrated into the farming system, communal pastures were developed. If this system is properly managed, then the use of pasture land and forest belts enhances and maintains the overall productivity of the fertile core plots.

Colonial governments repeatedly made the mistake of assuming that pastures (*rweya*) were simply unused arable land, but experiments meant to "make the *rweya* bloom", all ended in dismal failure.

The land use pattern is fragile, and if it is pushed beyond certain limits the ecosystem begins to decline. McMaster (1960, pp.82-83) believed that the very successes of the Karagwe kingdom in raising large cattle herds led to "lowered soil fertility" as "the climax vegetation was ravaged and the grasslands and open vegetation multiplied". The *rweya* pasture lands have a limited capacity that can be damaged by overgrazing. Cattle, naturally, eat the more nutritious grasses first, leaving only the coarser grasses. The Haya often burn this coarse grass to allow better varieties to grow, but if the herds are allowed to eat the new tender shoots, then the laterite soils dry out, which leads to a vicious circle of environmental degradation.

The quality of pasture land on the densely populated coastal ridges is declining as a result of such overgrazing. Once the soil becomes "laterized" then only the less nutritious grasses can survive, such as *Digitarium, Eragrostis, Hyparrhenia, Loudetia, and Brachiaria* (Rald and Rald, 1975, p.51). As a result, the stock-carrying capacity of the grasslands is low and has reached a critical level in areas where the communal pasture is overgrazed; these poorly nourished cows are capable of producing only about one-quarter to one-third of their milk capacity (Tibaijuka, 1979, p.55; Friedrich, 1968; Rald and Rald, 1975).

As population increases, people are forced to farm land that they previously considered marginal. This reduces available grassland while at the same time more cows are added to the communal herd and forests are cut down for firewood. These competing land uses lead to intensified leaching and laterization of the originally poorest soils in the region.

Pasture (*rweya*) land is used for planting some annual crops as well as grazing. But the women must construct tiny plots, *omusiri,* by transporting earth to mounds where the crops are cultivated; these require fallow periods of eight to twelve years, if these sites are to remain sustainable. However, in a situation of increasing population pressures, as is now the case in the Bukoba and Muleba Districts, fallow is reduced to anywhere from three to seven years and the *omusiri* land is becoming degraded (Tibaijuka, 1979, p.30).

The major advantage of the traditional cattle grazing system is its efficient use of labour. Each "customary village", has an *omukondo* or cattle owners association with twenty or so member households, a set pasture area, and a herd manager. (Today's official villages are administrative units and are usually an amalgamation of three or four customary villages.) The herd manager or *mukondo* decides on the area to be used for each day's grazing, and checks the herd at the end of the day. Children bring the family cattle to the appointed area each morning, and a designated representative, *omulisa* herds the cattle for the day. The optimum herd size is never more than fifty or sixty. With the regional average of four head of cattle per owner, each *omulisa* only needs to put in one day of herding every fortnight or so. The traditional cattle are rarely stall-fed and since they are grazed only five to eight hours per day they tend to be undernourished and, as noted, poor milk producers.

Richer farmers with larger herds are excluded from this system: they must either hire full-time or part-time cattle attendants. (A full-time "cowboy" is paid a maximum Tanzanian shilling equivalent of twenty dollars per month plus food, a part-timer gets one dollar per day.)

The government, with technical and financial assistance from the Netherlands, has devised a popular and ecologically-sound supplement to traditional Haya animal husbandry. The project runs breeding ranches that raise specially bred dairy cattle: these cows are sold to farmers on condition that they will be stall fed and kept away from the communal herds in order to help prevent dairy cows from picking up diseases. This labour-intensive, zero-grazing approach helps to employ surplus male population, who are hired to cultivate and transport a rich broadleaf called "Guatemala grass" or "Napier grass". It is cultivated beside roads, on the edge of cliffs, in swamps, ravines, and other places where cattle could not normally graze. Another ecologically attractive offshoot other project is that cattle are fed diet supplements of cotton silicate and waste molasses, thereby using up by-products of the Chato cotton ginnery and the Kagera sugar estate.

Zero-grazing relieves pressure on the pasture areas and the purebred dairy cows provide a much larger output of milk (needed by children) and manure (needed by the banana plantations). The biggest problem with this system is that the initial cost of the cows and shed is beyond the means of all but the richest farmers.

A possible solution to the problem of pasture degradation would be to reorganize the traditional cattle owners *kyama,* making it easier for poorer farmers to purchase dairy cows on a collective basis. Trusted community workers from the cooperatives or church sponsored non-governmental organizations could help and encourage the village *omukondo* to exchange traditional cows for dairy cows (on condition that villagers replant the pastures with more nutritious grasses and reduce grazing). This type of project has the additional advantage of employing surplus male labour and encouraging community initiative by helping to restore some of the cooperative spirit that has been eroded in the past few decades.

Insects and Other Pests

Population growth increases the risks of pest infestation in the banana groves of Bukoba and Muleba districts where the spread of banana weevils (*Cosmopolite sordidus*) and nematodes (predominantly *Pratylenchus goodey*) is endemic. These pests reduce crop yields from twenty to ninety-five percent (Tibaijuka, 1984, p.116).

The recommended control was *dieldrien*, an insecticide used extensively throughout East Africa, but in the past fifteen years weevils have developed resistance to this cylodiene. Pesticide use is unregulated and most interviewees cited examples of farmers who killed their crops through improper use of insecticide or herbicide. It was not until 1973 that researchers discovered that much of the crop damage attributed to weevils was actually caused by nematodes. In 1990 the Ministry of Agriculture recommended *Furadan;* however, this insecticide is costly, not readily available, and like all inorganic chemical amendments is likely to have damaging long run environmental side-effects. Insecticide users reported many problems growing beans and agricultural testing confirmed that the population of bean pests such as millipedes and snails increased rapidly when farmers used banana pesticides. More ecologically sustainable methods of controlling banana weevils have succeeded, such as raising chickens (or occasionally ducks) in enclosed banana plots, where the fowl will eat most of the pests. Unfortunately, peasants have such small cash flow that the high cost of fencing material and supplemental feed makes raising chickens as expensive as using insecticides and therefore beyond the means of most smallholders. In the past, farmers solved the pest problem by migrating to new plots, but this solution is no longer viable due to the acute land shortage in the region.

Paradoxically, banana yields are lowest in areas where farmers have the best incentives to improve yields. Farms nearest the urban markets where demand and prices are the highest are often the least productive. This is proof that a higher population density and intensive cropping lead to environmental damage, lower fertility and pest infestations (Tibaijuka, 1984, p.89). Small hold units are generally more productive in remoter less-populated districts, even though the farmers of the populated zones have much more available capital. Tibaijuka (1984) studied ten villages and she discovered that Rushaka was the one with the best marketing position and highest average prices; it is only sixteen kilometres from Bukoba, the largest food market in the region. Its marketing advantages were offset by its poor harvests as this village had the lowest yields of bananas per hectare.

Extension Services

In theory, the extension workers sent out by the Tanzanian government help the farmer to achieve greater productivity and eliminate problems such as pests. In practice, extension workers reach fewer than one-fifth of farmers; the situation has not improved in thirty years.

Rald's (1969) study of 52 Bukoba Rural District farmers, reported that 20% met with extension officers and 25% read farming news. Tibaijuka (1979) interviewed 50 Bukoba Rural District farmers, of whom 18% met with extension workers, 10% read *Ukulima wa Kisasa* (Better Farming) the Ministry newsletter, and 8% listened to farm news on the radio. My study, based on 250 life history interviews and supplements, found that: 11% read farm improvement literature, 7% listened to farm reports on radio, and 15% met with extension workers. (Many farmers complained that they could not listen to the radio farm broadcasts because they were at 10 AM when they were out working in the fields.) My most encouraging result was that 23% of farmers had either taken courses in improved farming or attended a supervised tour of a demonstration plot.

Tibaijuka (1979 p.70) explained why extension services have a limited impact, they do not have enough personnel and adequate transport to reach all the farmers. During the period of fieldwork I met several extension workers who presented their own set of constraints and complaints. For example, Julius Mugambwa is an extension officer in Muleba District: he is one of the luckier ones because he has a motorcycle at his disposal. Nevertheless, he was unable to visit all the farmers under his jurisdiction: he quite candidly told me that he usually only helped the richer more progressive farmers — those most likely to carry out his recommendations. Selection is the rule both in the offer and the acceptance of advice. Nagiza Rugensha, a woman agricultural officer, could only reach limited numbers of clients because she had to travel on foot. She reported that villagers often would not accept her advice both because she is a woman, and because she had not yet been able to acquire her own land. Farmers often told her, "How can you advise us when you don't have your own farm".

Under these conditions, it seems very likely that extension services will continue to target the relatively well-to-do. In my sample, the use of extension services is heavily concentrated among the owners of dairy cows, who protect their investment with mandatory cattle dips and frequent consultations with veterinary officers.

5 AIDS and Depopulation

The last chapter highlighted how the Haya use the regional resource base to sustain their farming system and culture. Nevertheless, rapid population growth threatens the long established pattern of a people living in harmony with their natural resource base. Rapid population growth is a stress on the agricultural and social system and too often threatens the health of women forced by patriarchal custom to have too many babies. Unfortunately the strongest factor working against the local population explosion is mortality: the AIDS epidemic as a crisis; a severe demographic threat that has already changed many economic and cultural practices. In order to understand the nature of this threat one must first consider the demography of the region and recent changes.

Haya Demography

The British Colonial Administration for East Africa carried out its first real census (based on an actual enumeration) in 1948. Since then the official population count for the Kagera region has been as follows:

- 1948= 300,000
- 1957= 368,000
- 1967= 659,000
- 1978= 1,010,000
- 1988= 1,326,000

Population growth in the longer term depends on subsistence agriculture and potential alternative incomes, what (Allan 1965) calls "the human carrying capacity". The large increase in population between 1957 and 1988 is attributable to two primary factors; increasing fertility and declining mortality. The decline in mortality was a direct result of improvements in the primary health care system in Tanzania. One expert (Vaughan 1986, p.160) estimates that in most traditional African societies about one-half of the children died of diseases ranging from the exotic (Lassa fever, green monkey fever) to the major killers (malaria, estimated to kill one million African children annually, diarrhea, pneumonia, measles, diphtheria). When there are shortfalls in food production, children are not fed properly and therefore the effects of childhood diseases become more dangerous and deadly (Sai 1986, p.139).

The Tanzanian national census figures are not entirely reliable but they indicate a trend toward recent reductions in infant mortality: 190 per thousand in 1957, 145 per thousand in 1967, 135 per thousand in 1978, 130 per thousand in 1988. In 1967, a Tanzanian national survey reported that 740 children per thousand lived until the crucial threshold age of five — this is twice as many as expected in a traditional society with no medical intervention (Koponen 1986, p.41).

Koponen argues that rising fertility in the twentieth century is a more important population growth factor than declining child mortality. In pre-colonial and early colonial times "postpartum non-susceptibility" was the rule: women breastfed their children for at least two years. Child spacing reduces births because of postpartum anovulation, a nursing woman is rarely fertile, and also because of strict rules of post-partum sexual abstinence during nursing. If these failed herbal and mechanical abortion methods were available. The end result was a "spacing of several years between births". In the last two decades, child spacing norms have been eroded by the "economic, social, political and cultural changes which began with colonialism" (Koponen 1986, pp.42-49; Dawson, 1987; O'Brien, 1987; Piché, 1987).

One such change has been the influx of sexually transmitted diseases. In colonial times this region was plagued by syphilis. At least two experts, Lutheran Bishop Kabira and medical anthropologist Priscilla Reining attributed the low birth rates of the 1950s to sterility resulting from syphilis and gonorrhoea. In the period before antibiotics syphilis often caused sterility. Women who feared sterility would not practice child spacing but would prefer to have as many children as quickly as possible. Haya culture rewards fecundity and sanctions infecundity and even the advent of Christianity has reinforced this pattern since the majority of Haya are Catholic. Economic factors also come into play when unmarried women give birth to children in rapid succession because they believe that they will receive child support from a partner or succession of partners.

Flora is a divorced woman in her mid thirties. She already has seven children from three different husbands. She receives about one-quarter of her income in the form of remittances from ex-spouses. She farms a small plot of land and supplements farm and remittance income by working as a midwife and traditional healer. One of her sons suffers from a disability and requires extra care. Needless to say Flora is constantly working to support her family.

The high rate of population growth is also related to the age structure of the population. As in most other areas of Africa, over one-half of the people in Kagera region are under fifteen years. This leads to rapid growth caused by "population momentum", a term that demographers use when "the next generation of parents has already been born". Anna Tibaijuka (1979, p.110) projected that Kagera's population would surpass two million by the year 2000; her prediction has come to pass but in a way she did not foresee. After President Habyarimana was assassinated in 1994 and the ensuing Rwandan civil war, about one-half million Rwandese refugees fled across the Tanzanian border to this region. Kagera's population in 1996 reached the two million level but refugees were fed by the World Food Programme and therefore constituted a separate category directly dependent on outside agencies. Most were repatriated back to Rwanda, see chapter six.

Demographic Structures and Gender Constraints

As we have seen, Kagera's population tripled over forty years. Population growth in an agriculturally-based society is contingent on the carrying capacity of its agricultural land (Allan 1965). In Kagera Region, studies carried out in the 1930s demonstrated that the plantain/coffee economy was supporting a population density of 700/sq. km. of improved arable land (Milne 1938). More recent studies delimit the same maximum population density, one that can not be exceeded given the existing farm technology (McMaster, 1968; Reining 1962; Freidrich, 1968; Boesen, Madsen and Moody, 1977; Schmidt, 1979).

Since the population density on the arable land cannot increase, and there is no evidence of massive urban growth, then the only reasonable explanation for the tripling of the census population is that the farming system expanded onto lands which had previously been considered marginal. Soils need to be upgraded, land needs to be cleared, new plantings undertaken, and the location is not as accessible to markets. Out-migration was another crucial adaptation to population pressure. The two processes were complementary since the most common method to accumulate savings, both for land purchase and development, was to work off-the-farm for several years.

Sampled households reported that 60 percent of sons and daughters between the ages of 25 and 45 migrated to urban areas or to full-time non-farm employment for an average of 13 years; of these, 39 percent are female. Prior to Independence and the establishment of a national education system, missionaries set up an extensive network of schools in the region (then called Bukoba). Thus, the population is relatively well educated. Most parents in the sample encourage their children to further their education and they make great sacrifices to that end. As a result, a high proportion of the migrant youth are employed in high-status jobs. Mbilinyi (1976) believes that the Haya occupy a greater number of high level posts than any other Tanzanian ethnic group. In our sample of two hundred and fifty households, ten had sent children to university, four outside the country: this is about twenty times greater than the national average.

The end result of out-migration is that those who actually own the bulk of the arable land are past the productive peak for manual farm labour; the sample median age for head of household is over forty. According to the 1978 census, 49 percent of the inhabitants of the region were under 15 years of age, and 14 percent were over 55. Therefore, only one-third of the villagers are in the prime of working life for hoe agriculture.

Out-migration from heavily-settled areas of Kagera, such as Bukoba and Muleba districts, will increase unless alternatives such as population control, rural industries, or intensive farming programs are provided. Since over 60 percent of the village residents are under 25 years of age — and the average sampled household consists of 5.2 persons — the population could double in fifteen or twenty years. The existing farming system could not produce enough food to sustain this population.

Out-migration has augmented the number of female-headed households. In some cases, men who have divorced their wives leave these women with occupancy rights to a portion of the land so that the ex-wives can grow food and provide subsistence to the children. In other cases divorced women cannot survive unless they migrate and leave the farms they have worked hard to develop.

Out-migration

Out-migration is a decisive determinant of population growth. Bukoba town has never attracted massive population migrations, in the manner of Mwanza and Dar es Salaam. Unlike the migrants of West and Southern Africa who tend to reside permanently in urban centres, the Haya more often migrate for a period of years but then return. Out-migration and the development of new farm land are complementary processes since for many Haya the most effective method to accumulate savings, both for the purchase and development of arable land, is to work off-the-farm for several years. Out-migrants are of two types: 1) those with the best education and training able to find good jobs in the public sector or start successful private businesses — the most successful of these migrate to the commercial centres, Dar es Salaam, Arusha, Mwanza; 2) those who are landless and poor (usually badly educated and more often women) who migrate merely to survive. Sometimes even a small amount of capital can be wisely invested and lead to relative security when migrants return.

Paulo studied accounting in Dar es Salaam and then was awarded a scholarship to study cooperative financing in Canada. He taught in a Tanzanian college until he decided to take early retirement at age fifty. Teacher salaries are low in Tanzania so he did not return with a very large amount of capital, but he was saving money by living in a mud house without many amenities and reinvesting in building up his core plot and an additional tree farm. Using part of his core plot as a nursery and harvesting trees for firewood provided him with a high income by village standards.

As noted, 39 percent of sampled out migrants are female (of whom two-fifths were government employees, teachers and nurses). The rest are homemakers, service workers or students. Haya women practising prostitution became notorious in the Pumwani district of Nairobi during the colonial period (White, 1990). Successful Haya prostitutes generally return home to set up independent households: as independent farmers they are accorded high status and become female role models. Haya prostitutes are omnipresent in all the urban centres of East Africa. Bader (1975) and Swantz (1986) both present cases of Haya women who returned to purchase farms with capital saved from urban prostitution. As Bader (1975) points out, many women out migrants who return to set up farms are considered to be returned prostitutes. In a later chapter I present two case histories of women who became rich farmers after accumulating money from prostitution. Paradoxically, successful prostitutes may be at lower risk from AIDS than village women because they understand the disease better and have access to condoms.

Urbanization and the Rwanda Crisis

The official regional urban population is about ten percent of the total. Tibaijuka (1979, p.100) estimates that no more than two percent are "fully urban", by which she means they do not control even a small farm plot. Bukoba town has a population of 49,000 (1988 census) and is the fastest growing town in the region, Muleba, has about ten thousand. Other urban centres are either in decline (as in the case of Kamachumu, formerly a trading centre) or very small.

There was an small influx of Rwandese and Burundi refugees in the 1970s and 1980s, but as noted, in 1994 about one-half million Rwandese fled to the Tanzanian side of the border to refugee camps administered by the UNHCR. Very few Rwandans were allowed to remain as long-term Tanzanian residents after the mass repatriations of 1996. Given the numbers of refugees and high population density among the Haya, there was little chance that one-half million refugees could have become permanent self supporting settlers even in the more sparsely settled border region.

The AIDS Crisis

Since the mid-1980s Kagera region (due to its proximity to Uganda and central Africa) has the reputation for having a higher rate of deaths from AIDS than any other region of Tanzania. Reliable statistics are not available but during the period between February 1987 and July 1989 I assembled evidence that does justify the prevalent crisis mentality. Dr. Twagiraresu Medical Director of the Ndolage Lutheran hospital reported that in 1984 there were less than 100 HIV positive diagnoses, just over 200 in 1985 and about 200 in the first five months of 1986. Since then, the incidence of diagnosed cases dropped drastically, mainly because AIDS victims no longer come to hospital since no cure is available and the family or clan must pay for expensive arrangements to transport the bodies of the deceased back home.

At New Years, the Catholic Church sponsors a song competition to 'divine' the theme of the year. In 1987 the year of *wemoge*, the reform of habits, this song won a prize.

> This is the year to reform
> be careful about AIDS
> fat girls and fat boys don't trust them
> they can have AIDS also.
> Fat girl, fat boy leave me alone.

In 1988 the "Stereo Music House" in Bukoba recorded a cassette of traditional style songs. The first song on the cassette *Kanana* (life) contains these words

> Life is life, go and find your lover,

Leave my lover he is sophisticated
Don't come to me I am very afraid of the AIDS virus.

Propaganda songs are not the only expression of popular concern. In September and October 1988 we studied twenty unofficial bars in a village noted for illegal distilling, described in a later chapter. AIDS, was the most frequently discussed topic. In this village of about two thousand people there were five funerals for AIDS victims within three weeks. In three sample villages (with a combined residential population of about 5000) we recorded forty-two funerals of reputed AIDS victims in the last quarter of 1988. This number is somewhat misleading since some of these funerals were for urban migrants whose bodies were shipped back for burial.

The 250 heads of household interviewed, reported deaths in the immediate family (children, parents, brothers and sisters) in the past year. The average number of deaths is 0.8: AIDS accounted for a larger number than any other cause — 38 percent of the deaths; malaria, was second at 20 percent. It is impossible make any precise assessment of the extent of the AIDS crisis for a number of reasons: people's unwillingness to report AIDS deaths, inaccuracies in diagnoses, limited sampling possibilities, geographic mobility, and so on. Nevertheless, in the 1987-89 period, AIDS appeared to be the largest single cause of death for Kagera region residents.

Another indication of the seriousness of the problem is the change in funeral customs. Because of the large number of deaths many Haya stopped bringing the bereaved family or neighbours the traditional gifts of bunch of bananas or a "pound note" (formerly twenty, now ten times that much or more depending on current exchange rates). Many mourners in the 1990s brought only token gifts or none at all. Until very recently it was the custom for most people in a village to take two or three days off from farm work to attend funerals. Now many people only visit funerals for a few minutes at a time or not at all, and we frequently heard the complaint that if everyone attended all funerals in the culturally sanctioned manner, no one would get any work done. Since most AIDS victims are in their reproductive years, the disease may replicate the infecundity problem of the 1950s or if the worst predictions come true even lead to depopulation, although many who die of AIDS have already had children.

According to PARTAGE, an NGO centred in France, there were approximately 25,000 orphans in Kagera Region whose parents have died of AIDS, personal communication Jackson Bube, 1991. The extended family and clan system cannot absorb this many children, which is why PARTAGE is involved in assisting orphans whose relatives are not willing or able to adopt them.

Summary Prognosis

It is reasonable to assume that the tripling of the census population is the result of high rates of fertility leading to population increases among the Haya. To date neither the AIDS crisis nor malaria nor temporary out-migration has meant a decline in population. On the contrary the AIDS crisis is merely another very strong factor influencing fertile women to reduce child-spacing and to have as many children as possible. While it is still too early to predict the end result of the AIDS epidemic, it seems likely that AIDS will not lead to significant depopulation.

The regional farming system has remained stable for at least fifty years: in 1938, Milne demonstrated that the plantain/coffee shamba economy could support a maximum density of 720 people for every square kilometre of improved arable land (Milne, 1938). More recent studies (McMaster,1960; Reining, 1967; Friedrich, 1968; Boesen,1977; Schmidt, 1979) delimit a similar population density that can be taken as a maximum. A Haya farm village cannot support more than about 700 people per square kilometre. In the 1990s, Haya survival required that the farming system expand onto lands which had previously been considered marginal. In subsistence agriculture the boundary between clearly cultivable and uncultivable land depends on access to resources. Marginal social groups such as women heads of household or landless youths may be forced onto land in poor locations, with shallow or unfertile soils because that is all they can afford. Even with extra effort their survival is precarious.

The minimum land requirement of an average household of six

is a *permanent plot* of one acre or 0.4 hectares (Tibaijuka 1979, p.23). Rald (1969, p.28) reported that forty per cent of the core plots in the inner districts were below this minimum and Tibaijuka (1979, pp.19, 23) found that the per capita household farm area (including fallow areas) declined from 1.8 hectares in 1967, to 1.2 ha. in 1979. During my fieldwork period, the average landholding of my respondents was still 1.2 hectares: this indicates that virtually all the arable land in the Bukoba and Muleba Districts was occupied and that population is at its maximum density.

Since even marginal land is now occupied, new survival strategies will be necessary. Survival will require a greater degree of trade and industry in the informal economy. The AIDS crisis is terrible for those families involved in personal tragedy, but its effects have not yet had any impact on the village landholding patterns or regional demographics.

6 Rwandese Refugees in Tanzania

The influx of close to one million refugees into North-Western Tanzania constituted a threat to the ecosystem, the infrastructure, the culture, and above all to security. The Tanzanian government and most citizens wanted the refugees out, but few refugees were willing to *voluntarily* repatriate; however at the end of 1996, services in the camps were withdrawn and refugees were given no choice. This was done because Kagera region was subjected to stresses that threatened both regional and national security and could have forestalled both large and small-scale development initiatives. Tanzania faced a very serious dilemma; there was no way to sustainably settle so many refugees so the government took the risk that *refoulement* would not lead to either international censure or a cross border guerilla war.

Power Politics

Tanzania is a signatory to both the Geneva Convention (The Convention Relating to the Status of Refugees 1951) and the more liberal 1969, Organization for African Unity, Convention Governing the Specific Aspects of Refugees in Africa. These international protocols specify how refugees are *supposed to be treated.* Both these treaty regimes forbid *refoulement,* involuntary repatriation, however, this has not prevented periodic expulsion of refugees from African nations including Tanzania, (The Lawyers Committee for Human Rights, 1995). President Mobutu of Zaire, in August 1995, began forcing Rwandan refugees to repatriate to Rwanda. This was done as a protest against the United Nations lifting the ban on arms shipments to Rwanda. Refugees were pivotal players in regional power politics. The Democratic republic of the Congo (DRC formerly Zaire), Kenya, and to a lesser degree Tanzania fear the rearmed

Rwandan Patriotic Front (RPF) government because in their estimation, it maintains too close ties to Uganda. A Rwanda/Uganda axis could block the trade routes between the Great Lakes into Central Africa, and a combined force of these two fierce armies, experienced in successful guerrilla warfare, could pose a threat to its neighbours; the Tutsi led rebellion in DRC, Democratic Republic of the Congo continues to destabilize the largest country in the region. Pervasive insecurity rules the Great Lakes Region with twenty armed forces participating in hostilities; these include armed Hutu supporting Kabila, along with troops from Zimbabwe, Angola, Namibia and Chad, while Rwandan, Burundi, and Ugandan troops are supporting rebel forces, *The East African,* Nov 30-Dec 6, 1998. Uganda, despite strong leadership and a growth rate of seven percent in 1996 and 1997, has not yet made peace within its borders and on the frontiers. It faces both internal insurrectionary forces such as the Lords Resistance Army led by Joseph Kony near the Sudan border and the Allied Democratic Front in the West, as well as external raiders such as rebel *interahamwe* and Congolese in the Virunga forests along the Western border.

Hutu extremists naturally feared *refoulement*, however some might have hoped for it in order to boost the rebel forces in exile on the Congo front and perhaps to be able to perpetrate acts of sabotage and terrorism inside Rwanda if the pro-Kabila forces prevail and the Hutu army in exile is granted asylum in the DRC.

Despite combatants on its frontiers the RPF loses international credibility every time repatriated refugees are attacked. Such highly visible abuses of human rights could jeopardize current and future international assistance and help destabilize the country in the long term.

Warning Signals: Was a Crisis Avoided?

Stewart Wallis, overseas director of Oxfam, warned of an impending crisis in 1995:

> We believe there is the threat of a major potential catastrophe... moreover there is a growing sense in the region that we are sitting

> on a powder keg and the food crisis could be what ignites it. (*Daily Nation*, Nairobi, March 17, 1995)

In January 1995 I worked as a volunteer assisting with food distributions in the Tanzanian camps. Each week we doled out the ration to each registered refugee consisting of three items: a three litre measure of dry maize (corn kernels) a half litre scoop of dried beans and a three hundred millilitre stock of corn/soy flour. When the authorities reduced this survival level allocation in 1996, refugees had a choice of repatriation or facing hunger, malnutrition and eventually starvation. The Mauritanian government in 1989 used the same method of "forcible repatriation by the back door" (Lawyers Committee for Human Rights, 1995).

From Hospitality to Hostility

Tom Kuhlman (1994, p.135) concluded that Tanzania was the most hospitable African country, because of the "the generosity with which land was made available" and "the way the Tanzanian Government treated refugees equally with nationals". When the number of refugees from Rwanda and Burundi were in thousands they were given land and welcomed to the neighbouring Kagera region, but after almost one million Rwandese exiles entered Tanzania, the Government's hospitality turned into hostility. At the end of 1994, Mr Patrick Chokala, then Press Secretary to former President Mwinyi stated:

> (Rwandese) refugees could not involve themselves in any income generating activity hence indulged themselves in criminal activities outside the camps . . . creating enmity between the refugees and residents of the areas. The Government is taking measures to return Rwandese refugees to their country . . .

Former Presidents Mwinyi of Tanzania and Mobutu of Zaire resolved to find a solution to the problem by establishing camps in safe zones in Rwanda under the surveillance of the OAU and UN (*Daily News*, Dar es

Salaam, December 24, 1994). This proposal was based on a joint communiqué by Presidents Mwinyi and Mobutu issued in Gabadolite Zaire on December 20, 1994. President Mwinyi had proposed another resettlement option in early December — that Rwandese be moved to camps in Southern Tanzania left vacant by refugees from Mozambique (*The East African*, Nairobi, December 26, 1994).

In January 1995, the Presidents of Tanzania, Kenya, Uganda, Burundi, Zambia, and the Prime Minister of Zaire met in Nairobi for a Regional Summit. The agenda focused on how to repatriate Rwandese refugees, intimidated by criminals in the camps. One report stated that "most fugitives in Zaire, Tanzania and Burundi are staying put either because they are guilty or intimidated by adherents of the previous regime" (*The East African*, Nairobi, April 3-6, 1995). At the January 1995 summit even the Rwandan delegates were obsessed with the issue of intimidation inside the camps, their main concern was that Tanzania and Zaire were granting sanctuary to men who were training to destabilize the Kigali Government (*The East African,* January 9-15, 1995 and January 23-29, 1995).

These leaders refused to acknowledge that Rwandese refugees were afraid to repatriate because of dangers on the journey or fear of retaliation at home such as the slaughter of at least 2000 refugees in Rwanda in April 1995 (*The New York Times)* May 21, 1995. This attack took place shortly after Tanzanian officials closed the border with Burundi and blocked over 50,000 Rwandan Hutu refugees fleeing camps in Burundi.

> The Prime Minister of Tanzania, Mr. Cleopa Msuya stated: We are overwhelmed by the number of refugees which is straining our resources and undermining our security. Tanzania is unable to receive any more refugees and it is up to Rwanda to give guarantees regarding their (refugees) security. (*Daily Nation*, Nairobi, April 3, 1995)

Refugees and Regional Development; Costs and Benefits

With the wisdom of hindsight it is obvious that the Tanzanian Government clearly believed that the costs of maintaining three-quarters of a million refugees was greater than any benefits. However the issue was clouded by many prejudices and fears based on breaches of security. When I visited Benaco in July 1994, it was the largest refugee camp in the world with about 250,000 residents. In the 1980s the Kagera region of Tanzania was a neglected and peripheral zone and the refugee-receiving areas, the Ngara and Karagwe districts, along the Rwanda frontier, were its least developed portions. In order to support the massive relief organization the UNHCR and the NGOs financed some road works, improved water systems and assembled a radio communications network.

My first impression of Benaco camp and its environs was a beehive of activity, a boom town. The small unpaved airstrip was overrun by small planes chartered by the UNHCR, the Red Cross, AMREF, and other relief organizations. Convoys of expensive new four wheel drive vehicles and larger UNHCR and WFP lorries sped by as they shuttled from the camp to the airstrip and to the staff housing compounds and the Tanzanian municipality of Ngara.

A minority of local farmers were getting rich by selling food and other supplies to local camps. The camps were also a boom to residents of nearby villages, hired as drivers, cooks, cleaners, security guards and construction workers. Since the UNHCR and the Red Cross payed the highest salaries, they were the first choice for workers. Those unable to find work at the top of the scale would then try the less-desirable NGOs paying lower salaries. For local residents this employment was a much welcomed source of cash because before the relief effort there was almost no salaried work available in these districts except in the informal sector at rates below the official minimum wage of about one US dollar per day for casual labour.

In January 1995 when I returned to Kagera region my impressions were different, it was clear that the honeymoon was over. The Tanzanian Government and other organizations commissioned several environmental assessments. I am aware of three reports, one by the Ngara

Development Trust Fund (NGADETO), another by Professor Reginald Green of the Institute of Development Studies, Sussex, another by Dr. Richard Black also of Sussex. These experts all stressed the problems of environmental degradation, extra wear and tear on the already weak infrastructure, and social problems such as tensions between the host and refugee population.

Environmental Damage

The NGADETO report tried to quantify some of the environmental stresses. For example they estimated that refugees consumed one million litres of water per day. The water level in the lakes and rivers adjoining the camps had noticeably gone down and most above-ground water supplies were polluted in the vicinity of camps (*The East African*, December 26, 1994). Nature more than replenished the water table with massive flooding caused by unseasonal El Nino rains in late 1997 and early 1998.

Deforestation posed another critical problem. The NGADETO report claimed that refugees in Kagera used up 176,000 cubic metres of firewood in six months; this is the equivalent of 3,500 hectares of forest (*The East African* December 26, 1994). Professor Green claimed that some smallholders in villages adjacent to the refugee camps spent an extra two hours per day collecting firewood and drawing water because of resource depletion. Deforestation is definitely a problem; relief personnel marked some 450,000 trees, by painting white rings as a sign that they were not to be cut. Nevertheless UNHCR officials at Kagenyi, the camp I visited in January 1995, reported that illegal tree cutting was carried out up to fifty kilometres away from the camp.

Marking trees was an ineffective mode of protection since it was impossible to know whether or not wood entering the camps came from a marked tree. Firewood sellers predominated in the camp marketplaces and wood was the only product consistently available. The NGOs in the camps introduced a German Technical Assistance Agency initiative, the "GTZ oven" designed to *conserve* firewood. This circular clay structure was supposed to replace the traditional method of resting pots on three large stones: it requires far less wood. Yet in spite of promotional posters displayed around the camps, very few refugee householders adopted it. This may have been because there were not enough hands-on training courses demonstrating how to build these clay ovens. The traditional method is more adaptable and portable. Another reason for the lacklustre response to the GTZ ovens was insecurity; refugees were aware that they might be forced to leave at short notice— these fears were justified since this actually took place.

The deforestation problem continued to escalate and reforestation and large-scale tree planting projects are certainly necessary; in fact the problem has been compounded by the flooding and landslides caused by the torrential El Nino rains in late 1997 and early 1998. In a report three years before the floods, Professor Green estimated that it would require five years of replanting to restore wood supplies (*The East African*, January 9-15, 1995).

Destruction of Infrastructure

The refugee crisis diverted resources and increased wear and tear on the road system. The Green report highlighted how rehabilitation of the road network would cost over $15 million and take three years. It is not clear to what extent the relief effort *set back* road improvement since the roads of Kagera region have been in terrible condition for two decades. The UNHCR hired construction equipment from Cogefar, the Italian company building the tarmac road between Mwanza and the Rwanda border thus delaying that project.

Undoubtedly, the convoys of lorries and four wheel drive vehicles have increased the wear and tear on the road system but a host of international observers have also forced the local Tanzanian authorities to take their responsibilities seriously and to deploy road crews to do some maintenance work: international aid has also helped mobilize some resources. As Green noted "Tanzania has the personnel but not the money to do the job" (*The East African* January, 9-15, 1995). Donors who assist in road rehabilitation must ensure that adequate personnel and funds are assigned to maintenance and repair.

Education and health services were also affected by the relief effort. Schools and dispensaries were commandeered as temporary housing for refugees. This placed a heavy burden on already overextended facilities and the entry of an estimated 30,000 goats, cattle and other domestic animals brought in some animal infections (*The East African*, December 26, 1994). Clearly, schools and dispensaries need to be rebuilt and repaired and veterinary services need to be improved and extended in the Ngara and Karagwe districts; to do so the Tanzanian Government will require assistance.

Social Problems

The Green report identifies an additional problem; social and psychological stress resulting from proximity to the Rwandan catastrophe. Refugees were disorganized and violent in a way-inconceivable to Tanzanian villagers, similarly, being next door to a series of massacres, seeing bodies floating down the river Kagera and hearing shots was an upsetting situation (from an interview in *the East African* , December 26, 1994).

In the areas I visited, eighty per cent of refugees were women and children, many were suffering from post traumatic stress syndrome. Women and girls who were raped, feared unwanted pregnancies and exposure to HIV: they were the group most at risk to become psychologically traumatized (*The East African,* March 6-12, 1995). A press report illustrated this sort of trauma;

> ... a mother gave birth, gathered her strength for two hours then walked away from her infant, telling nurses "For nine months I have harboured an *interahamwe*, put it in prison or kill it whatever the government policy is".

The *interahamwe* was the Hutu civilian militia mainly responsible for the Tutsi genocide. The woman had been gang-raped by militiamen (*The East African,* April 3-9 1995).

Many Tanzanians feared that this traumatization and violence would spread beyond the camps: with so many refugees and so many former army and militia in residence.

Security Problems in the Camps

The biggest security problem in the camps *was* intimidation and assault. Tanzanian authorities were unable and perhaps unwilling to identify those responsible for the massacres and to bring them to trial. As a result

> ... the exiled political leadership, the militia or *interahamwe*, replicated the political and administrative structures that existed in Rwanda ... At Benaco, administrative power was in the hands of those responsible for the massacres in Rwanda. These structures were essentially conduits for extortion, killing, rape, and intimidation of aid workers. (The Lawyers Committee for Human Rights, 1995,pp.189-190)

When I visited Benaco camp in July, 1994, several aid workers reported attacks on huts and even hospitals as Rwandese were killed by other Rwandese. Ilsa, a Doctors Withhold Borders, MSF, volunteer from the Netherlands witnessed a revenge killing; a young man being hacked to death by a gang armed with *pangas* (machetes). She was in a car with another woman and was powerless to stop the killers. In September 1994, a Tanzanian driver was killed at Benaco, apparently a case of mistaken identity.

The Tanzanian authorities tightened security and posted more police to the areas adjoining the refugee camps. When I visited Kagenyi camp in January 1995 there were forty three police officers assigned to the camp, as opposed to five a month before. Yet even this number of police worked out to fewer than one officer for every thousand refugees.

Many refugees reported an increasing level of tension between the camps and the surrounding area. Theonius, the President of Rubwera Camp Refugees Association, reported that in January 1995 tensions were increasing between refugees and their Tanzanian hosts. The Tanzanian police assigned to crowd control duty tended to use a great deal of force, such as striking Rwandese with their batons, rather than just telling them to get back in the food distribution queue. The NGOs preferred to use their own staff or volunteer Rwandese security guards rather than call in the Tanzanian police.

The Tanzanian police rarely patrolled within the camps. They usually only entered the camps when called in to investigate or keep the peace, unfortunately, the internal Rwandan security system tended to be indistinguishable (or almost indistinguishable) from intimidation. Reports of "rampant looting" (*The East African,* December 26, 1994) might have been exaggerated but looting and theft in the areas around the camps was undoubtedly a problem, especially along the Rwanda/ Tanzania border where there are "ten times as many refugees as [Tanzanian] villagers" (*The East African*, January 9-15, 1995). A year after their mass arrival in May 1994, about one out of one thousand refugees were detained as criminals. Kagera regional commissioner Mr Mangala claimed that 606 Rwandan nationals and 77 Tanzanian accomplices were arrested and remanded in prison for armed robbery, general theft and rape (*The East African*, March 27-April 2, 1995). Many more such crimes must have gone "unremanded".

With the exception of Rwandan employees, very few camp personnel lived within the confines of the camps. In May and June 1994, the Red Cross, CONCERN and other relief organizations set up their compounds inside the refugee resettlement areas, wanting to be closer to their clients, but by the end of the year all the relief groups moved their camps at least ten kilometres away on adjoining ridges. Ingrid Jaeger, head of CONCERN at Kagenyi, claimed that in the old compound they were constantly besieged by refugees begging for food and clothing; it is CONCERN's mandate to distribute these. CONCERN was fearful that harassment might turn to aggression and they relocated their camp twenty kilometres away.

In early January 1995 I stayed at the Baptist Relief Services compound, the last NGO remaining inside the Kagenyi camp perimeter. As they made their final preparations to leave the refugee camp and relocate to a higher ridge, there were signs of unrest inside the camp. Between four and six in the morning we could hear patrols of about ten or fifteen joggers as they ran along the camp road outside the thatch partitions enclosing our compound. Relief workers at Kagenyi believed that these joggers were military recruits getting into shape for a new round of violence. During the day we sometimes saw one or two young men in track suits training on the roads and footpaths. At one thirty in the morning, I heard the sound of drums, loudspeakers, shouts and marching music, I could see flashing lights in the distance. Many relief workers and press reports claimed these joggers and party-goers are the camp militia; the people who maintain order by intimidation. Were they training to eventually help destabilize Rwanda? How many from Tanzania are fighting in the Congo?

When I visited Kagenyi in January 1995, it provided shelter for 45,000 refugees and therefore was small compared to Benaco, the camp with a quarter million people. I could hear children playing games singing and laughing until about 11 PM. This regime indicated a lack of parental control and discipline and reflected the shortage of schools.

The Extended Security Threat

Six well-educated young Rwandese refugees articulated their concerns: Francis, Martin, Anterre, Consolata, Theonius and Jean-Pierre were all afraid to repatriate not because they feared camp paramilitary but because they feared what might happen to them on the journey and what awaited them when they returned home. Their education was needed to rebuild Rwanda but they had also learned from the first round of genocide, that intellectuals were the first to be hunted down and killed. One can only hope that despite their fears, they arrived safely and are now reintegrating into Rwandan society.

The OAU Refugee Convention, Article II(6) requires that refugees be settled as far as possible from the frontier of their country of origin: yet Benaco and its sister camps Lumozi and Lukole, housed over three hundred thousand refugees less than twenty kilometres from the border. Karagwe district camps, Kagenyi I and II, were so close to border that residents of the camps could *see* the Rwandan Patriotic Army (RPA) camp in a grove of trees on the other side of the Kagera river. The camps could easily have been attacked by Rwandan forces. The main reason for housing refugees so close to the border was that the most plentiful available water west of Lake Victoria were the small lakes straddling the frontier.

Tanzanian authorities claimed that many refugees entered their country with arms and stolen property (*The East African*, December 24, 1994). Yet if the dozens of Rwandese refugees who I spoke to are typical and are to be believed, then few arms should have been brought in since all said they were thoroughly searched before they entered Tanzania. The number of armed Rwandese in Tanzania may have been small but they undoubtedly troubled the Tanzanian authorities and local police.

The exiles in former Zaire posed a much more serious security threat, one that is in 1998 still reducing security throughout the Great Lakes region. Human Rights Watch claimed that Hutu leaders in exile recruited an army of fifty thousand men (*The New York Times,* May 31, 1995). Hutu *interahamwe* at Goma Zaire *were* building up a large supply of arms: Rwandan officials at the Nairobi Summit in January 1995 demanded that the Mobutu government should have returned Rwandan state property such as reported military goods — five helicopters, two hundred armed vehicles and a six seater plane (*The East African*, January, 9-15, 1995). This disputed military property is part of a much larger quantity of hidden arms that former Rwandan army and militia brought with them into Zaire and arms shipments further increased the tension levels. According to *(The Daily Nation*, Nairobi, March 29, 1995) ten to twelve "medium-sized Ilyushin aircraft . . . have been landing in Zaire to supply arms to supporters of the former government in Rwanda". The Rwandan government purchased R-4 automatic rifles, ammunition, grenades and rocket launchers from ARMSCOR, the South African State Armaments Corporation, but this was done on a purely commercial basis (*The Daily News*, Dar es Salaam, July 2, 1994). South African officials admitted that weapons may have been shipped to Rwanda and Rwandan exiles despite the embargo *(The New York Times,* May 31, 1995).

The Rwandan government feared the military build-up in Congo as a potential invasion force and the pro-Kabila forces probably consider them as dangerous allies. A Rwandan diplomat, who did not wish to be identified, stated in the *New York Times*, May 21, 1995, "We are actually creating the conditions for the Hutu in Zaire to invade."

In fact the opposite has happened as Rwandan troops and Congolese ethnic Tutsi are invading the DRC and without help from Zimbabwe, Angola, Chad, Namibia. and Sudan, Kabila's regime might have fallen in late 1998 and is holding a tenuous grip on power. From a Rwandan perspective this is both a struggle for diamond generated wealth as well as a pre-emptive military strike aimed at the former Rwandan army in exile and Hutu hardliners. However the Hutu are still arming themselves and some observers believe that these illegal arms deals and military operations are being financed by Hutu international drug trafficking *(The East African* Nov 30-Dec 6, 1998).

Conditions in the Camps

Refugees elected to stay in the camps, despite their complaints about poor conditions, because of both intimidation within the camps and fear of the unknown outside the camps. One of the major problems was food: refugees often complained about the quantity and the quality of their rations and claimed there was widespread malnutrition at Kagenyi. Medicins Sans Frontiers, MSF, was responsible for monitoring and treating actual cases of malnutrition. They sent "visitors" around the camps to identify the malnourished, especially mothers and infants. MSF brought the malnourished to a feeding centre where they received special food and treatment. In severe cases patients were kept there for observation. Isabella, a public health nurse responsible for the Kagenyi feeding centre, reported that malnutrition was not a serious problem at least by the standards applied to refugee camps in Africa; they had identified only forty malnourished mothers and children out of forty five thousand. Whether or not refugee complaints of malnutrition were exaggerated, most refugees found the camp diet monotonous and strange since they were used to eating plantain as a staple rather than maize. Even though the refugees were dissatisfied with the food, they still chose to stay in the camps.

Boredom was another serious problem in the camps. There were always large groups of men congregating in the open space outside our compound at Kagenyi, as one francophone refugee put it "ils choment" (they are unemployed and have nothing to do). Many zone leaders and the President of the Kagenyi refugee association stated the same thing. Adults in the camps had given up careers or abandoned farms and now sat in limbo. They had only a few square meters to grow crops or work on basket-weaving or other handicrafts. Children and teenagers were forced to interrupt their schooling; after several months some schools were set up but there was a shortage of teachers and supplies and only a small minority of children were attending school and only for one or two hours per day. Eight months after the crisis began, Kagenyi was just beginning to set up schools: one of my jobs as a volunteer was building makeshift plywood blackboards.

Camp housing was minimal, refugees lived in huts they constructed out of mud, sun dried brick, wood poles and thatch. They used blue UNHCR plastic ground sheets for a roof. The huts varied in quality depending on the labour power and skill of the occupants but the standards were well below any internationally acceptable level of habitation. Why did refugees endure hunger, frequent rain, cold nights, poor housing, disrupted education, unemployment and boredom? The answer must be that they feared repatriation even more.

Fear of Repatriation

Rwandese in exile had good reason to fear repatriation even if they were innocent of crimes against humanity. Many faced incredible hardship leaving Rwanda and were afraid that to return could mean another traumatic journey. The rumour mills in the camps and radio propaganda also exaggerated the all too real dangers and hardships. The first organized group to repatriate from Zaire were attacked by the militia; five hostages were abducted and the group of three hundred was sent back. The attack on an internal refugee camp in April 1995 resulted in at least two thousand deaths at the hands of the RPA, Rwandan Patriotic Army, composed mainly of Tutsi warriors. Not only the army but even civilians posed a threat: fourteen people were reported to have been stoned to death returning to their home village near the city of Butare, *(The New York Times*, May 1, 1995).

Hutu extremists used propaganda as a means of intimidation, they circulated tracts such as *The Rwandese People Accuse* or *Kangura* which means "Awakener" printed in Kenya (*The Daily* Nation, Nairobi, March 29, 1995). These hate publications exacerbated tensions by trying to justify the anti-Tutsi genocide and calling for invasion, for example the March 1995 issue of *Kangura* called on Hutu to unite, capture Kigali and liberate our country (*The East African*, April 10-16, 1995). Many refugees in the camps had access to radios and listened to broadcasts that repeated propaganda and false information. For example when I was at Benaco, the UNHCR and NGOs tried to establish a census system requiring everyone in the camp area to wear a plastic bracelet identical to the ones used by hospitals. Refugees refused to wear the identifications because the radio propaganda reported that the bracelets left an invisible mark which would appear under special light and that the RPF would use these marks to identify and kill people.

Even without intimidation and propaganda, refugees had good reasons to fear repatriation. The Rwandan justice system was crippled by the Hutu-led murder of most lawyers, judges and magistrates. Thousands (reports claimed 70,000 in May 1996) have been arrested in Rwanda and await trial. Many detainees have not been officially charged with a crime and a substantial number may be in prison because someone is settling a personal score or because they are claiming property which is now being occupied by thousands of Tutsi who returned to the country after the war subsided in July 1994.

Raymond Bonner reported:

> ... some (Tutsi) who fled pogroms 30 years ago ... have taken over houses left by Hutu who fled last year. When the real owners return, squatters often falsely accuse them of complicity in the genocide and soldiers arrest them ... but few cases have been investigated, no legal counsel has been provided and there are no resources for either. (*The East African*, January 16-22, 1995)

Some 400 suspected ringleaders of the genocide have been identified by the international tribunal that began trials in the first half of 1996. The tribunal relies on Interpol to bring the organizers of the genocide to

Arusha, Tanzania *(The Daily Nation,* September 14, 1995). However, this expensive showplace trial has done little to heal the wounds or even tried more than a handful of perpetrators of crimes against humanity. Few camp ringleaders or those responsible for genocide have been identified or brought to justice after repatriation. They have either succeeded in concealing their identity from authorities or have already fled. Those who conspired and carried out genocide will need to be tried in Rwanda: but impartial trials will not be possible in the near future, (*The East African*, April 3-9, 1995).

Amnesty International accused the Patriotic Rwandan Army (RPA) the armed wing of the Rwandan Patriotic Front, of killing "hundreds and possibly thousands" of prisoners and unarmed civilians. According to Amnesty "the Tutsi-dominated RPA carried out indiscriminate revenge killing of unarmed Hutu civilians in areas where Tutsis had been massacred" (*The Daily Nation*, 21 October 1994). The RPF government has pledged that it will not tolerate revenge killings but it will be very difficult to control soldiers in their teens and twenties who have lost friends and loved ones during the Hutu dominated genocide. Up to one million people, as many as 90 per cent of Tutsis in Rwanda prior to April 1994, were murdered (*The East African*, April 3-9, 1995).

The Rwandan economy is in ruins; this poses yet another impediment to development. Foreign aid is limited, yet without aid will governance be possible? The government only paid civil servants, teachers and the army twice between July 1994 and April 1995. Until the economy is rebuilt, looting and extortion will be hard to control and the threat remains of a breakdown in administration.

Will Repatriation Work?

Tanzania requested that Rwanda sign an agreement to repatriate 750,000 refugees. *The East African* (April 3-9, 1995) claimed that Tanzania, Rwanda and the UNHCR denied that the agreement meant automatic "mass repatriation" but the Tanzanian representatives wanted to repatriate "a significant number" as soon as possible. By the end of 1996 repatriation was a *fait accompli*. Tanzanian security fears, environmental destruction and "donor fatigue" all expedited the repatriation process. The WFP needed to raise twenty million dollars per month to feed the refugees in Tanzania. Once the flow of cash dried up, forced repatriation had to occur whether or not the UNHCR and other international bodies were satisfied that safe passage to Rwanda and secure resettlement in Rwanda was assured. At least by early 1998 repatriation hasn't triggered another round of mass murder.

Before the refugees crisis can be solved, Rwanda must to restore peace, security and political capacity? The government claims to be committed to reconciliation, the rule of law and democratic governance. But can the Rwandan government control revenge killings? Donors cancelled $30 million US after the killings in April 1995 (*The New York Times*, May 21, 1995). Without development assistance, could Rwanda rebuild the shattered economy and restore civil society, a free press, voluntary associations, independent judiciary and all the other benchmarks of a democratic society?

Given the long history of conflict in the region, will lasting peace be possible? Burundi remains on the verge of civil war and tensions remain high on the frontiers of Rwanda, Burundi, Uganda, Tanzania and the DRC; peace in the region will continue to require a concerted long-term international response. UN peacekeepers have little credibility in Rwanda: the government wanted fewer peacekeepers because of the costs and because of the UN was totally ineffective during the genocide of 1994 when most peacekeepers were forced to withdraw (*The New York Times* June 8,1995). Would OAU, the Organisation of African Unity, military supervision would be more acceptable? I agree with The Lawyers Committee for Human Rights (1995, p.192) that the "idea of peace corridors and safety zones as a method of in-country protection has been discredited" and neither the UN or any other body can guarantee the massive and continuous military presence needed to ensure effective protection.

Prognosis for the Future

Barrington Moore's classic text *The Social Origin Of Dictatorship and*

Democracy provides a formula for democracy often quoted by specialists in African politics, for example, the influential works of Sandbrook (1993) or Huntington (1984). The formula "no bourgeois, no democracy" presupposes a split between the market and the state and between those who control political power and those who control the economy. Sandbrook (1993) describes Africa's elites as "rent seeking". In other words, economic power comes from what Hyden (1983) calls "clan politics". Government contracts, special licensing agreements, inside information, high level corruption and interference in business agreements all tie the economy to politics and bureaucracy. In Rwanda the power of the state was mortgaged to former President Habyarimana's clan in the north of the country. Could the new RPF government somehow avoid this sort of rent-seeking regionalism and clan politics (Newbury, 1995)?

Sandbrook (1993) believes that a vigilant pro-democratic civil society might be constructed from intermediary institutions that substitute for a bourgeoisie —they could balance the dictatorial powers of the state and its bureaucratic control over the economy. The primary intermediary institutions would be the voluntary associations and non-governmental organizations or NGOs. The NGOs can be an important and some times the most important agent in the transition to democracy, but, in Rwanda the NGO sector was given an overwhelming amount of power, in the absence of a reliable government. One hundred and forty international organizations controlled much more money than the government. The Rwandan government expressed its frustration when it began to expel some NGOs at the beginning of 1996. Permanent Secretary, Christine Nyinawumwani stated: "there is no hostility for the NGOs but they tend to thrive on chaos". *The East African,* April 3-9, 1995, described what Ms. Nyinawumwani might have been thinking of when she used the term thriving.

> [The reporter] . . . stood on a Kigali Street and counted vehicles (mostly brand new Pajeros, Nissan Patrols, Toyota Land Cruisers) ... NGO staffers are usually identified by their portable radios with stubby antenna, their gender (mostly male) colour (mostly white) and language (mostly English). Some are billeted in

> Kigali's $120 a-night Milles Collines Hotel but they tend to gather at the nearby American Club for meals.

The wide social, cultural and economic distance between the expatriate staff of NGOs and UN agencies and the Rwandan population is a symptom of a major structural problem, the NGOs have more resources at their disposal than the government. Rebecca Dodd, reported:

> .... the ruling Rwandan Patriotic Front (RPF) is livid... It has watched with gritted teeth as aid has poured into NGOs and various UN agencies but not into its own coffers ... It has been left unable to govern and the humanitarian lobby is running the country. Only a third of the civil service is in place and those who are working here have not been paid for three months .... there is a shortage of everything from telephones to envelopes. (*The Sunday News*, Dar es Salaam, December 24, 1994)

Is the power of the NGOs justified by the human rights abuses still going on in Rwanda? The World Bank and other donors will continue to monitor human rights abuses and will need guarantees that the justice system is working fairly before they will release large sums of money. Is this creating a vicious circle?

Effective government will require assistance, since there is so much to rebuild: the NGOs had an important role to play but they are not a substitute for the state. Sandbrook (1993) promoted a formula for democratization in which grassroots organization run by concerned citizens put pressure on governments in a manner similar to an independent economic elite. The more enlightened NGO staff recognize the need to build up local NGO capacity as well as to support the government's confidence-creating measures. Robert Malella of Oxfam, Rwanda, pointed to the need to involve more Rwandese:

> Oxfam is establishing links with local groups working on rehabilitation and with the needy, particularly women. It is a multi-layered programme aimed at keeping the level of expatriates low. "We want Rwandans to work with Rwandans".

(*The East African* April 3-9, 1995)

The rhetoric is laudable but when will the international NGOs transfer power to the local NGOs? Will the government, the NGOs and the donors be able to reach an agreement or will the government simply continue to expel all foreign NGOs? Will this further jeopardize it credibility? Is Rwanda governable? Can two million refugees be re-integrated into the national mosaic? Will the justice system be able to function impartially or even to function at all? Is there any chance that the Hutu armed forces in exile will be disarmed and demobilized and reintegrated into Rwanda or peacefully and prosperously settled in DRC?

There is no way that the region can evolve into a peace zone until fundamental structural problems are resolved. Has the repatriation solved the refugee crisis in North-Western Tanzania or merely put it on hold until the next round? If Rwanda can resolve its most pressing internal problems then sustainable and democratic governance may be possible. However, the necessary time, money and good faith are all scarce commodities in the Great Lakes zone of Africa. Developed nations must put pressure on national donor agencies to do their part to help solve these problems. It may be far easier to sink into donor fatigue than to promote development. Let us hope that the international community will not be a party to more crimes against humanity in Rwanda, Burundi, Congo or anywhere in the Great Lakes region.

7 The Staff of Life

Banana trees provide the Haya's staple food, alcoholic and non alcoholic beverages as well as by-products used for agriculture and craft production. Haya farmers grow beer or "harsh bananas" on land that cannot support plantain or sweet bananas. The land areas not suitable to banana cultivation are used as pasture; cattle manure is a crucial element in maintaining and sustaining the farming system and food security. In brief, the farming system evolved as a logical and practical response to regional ecology. This chapter delineates three historical epochs in the development of this farming system: 1) its evolution, 2) the colonial intervention, and 3) recent innovations.

Evolution of the Banana Staple

Banana cultivation has dominated the area's agriculture for two millennia — according to archaeological evidence from preserved pollen grains (Schmidt, 1979, p.24). About four hundred years ago, pastoralist clans (the Bahinda) established themselves as rulers and introduced cattle to the regional farming system. As farmers learned to integrate cattle-raising and banana culture, they created a mixed economy based on cattle and bananas and they expanded food production to its present capacity, which can support a maximum of 700 people per square kilometre. Anacleti and Nagala (1980) (sic) in their article on the "cattle complex" in this region, point out that cattle (which produce little milk and meat) are valued primarily for their manure — used to fertilize the banana groves. The banana is more than a food and a source of alcoholic beverages. All parts of the plant are integrated into Haya culture and everyday life. The banana and its by-products, supply a host of needs. The leaves are used as a platter for serving the traditional meal, they are rolled to make temporary glasses, they are makeshift umbrellas, temporary shelters for weddings and other public gatherings and window shutters during an especially heavy downpour. Leaves, are fed to goats or cut into strips and dried. The strips are woven into baskets, mats, and thatch for roofing. The "trunk" of the banana tree is really a layered stem, which is separated, dried, and used for: mulch, bags, bottle and calabash stoppers, toilet wipes, latrine wall thatching, woven ropes and cords, seals and joining gaskets for liquor making, packaging at the market, head protection rings for carrying heavy loads, and *emitwalo* or *entende* the large (twenty litre or so) storage bags for beans, hung from the rafters (Tibaijuka, 1979, p.62). The banana skins and rotten fruit: are ploughed under as green manures or fed to chickens and cattle. The Haya even use bananas to make children's toys.

Bananas as Food

All the Haya individuals whom I spoke to said they prefer bananas to all other available carbohydrate foods. There are thirty varieties of cultivated banana in the region but these can be classified into four major divisions:
1. *ebitoke* or plantain, a hard green banana which are generally boiled with beans, fish or meat, to make *matoke* the staple food, about 60% of the trees are of this type;
2. *embire* or *kishubi*, harsh bananas used in beer brewing, about 30% of the trees are of this type;
3. *obwise* or sweet bananas, the type commonly consumed as bananas in the industrialized countries: the Haya eat these for desserts or as a snack, about 7% of the trees are of this type;
4. *ekonjwa* or "bananas of the king", this type are roasted or fried, they command a premium price but are consumed infrequently, they make up most of the remaining few percent of the trees (Ngaiza, 1981, p.73; Tibaijuka, 1979, p.36).

Despite their preference for bananas, consumption of other carbohydrates has increased in response to overpopulation and declining per capita banana production. More farmers are planting maize and cassava as well as beer bananas and using incomes from beer, liquor and trade to buy plantain: if they cannot afford bananas they buy cheaper foods. Part of my research was a food survey: the results demonstrate how food production and distribution operate in the 1990s. *Ebitoke*, was the main course in 49% of 2998 main (afternoon and evening) meals recorded by a randomly selected sample of ten families in three villages. It seems that the Haya usually get over one-half their daily caloric intake from bananas. (It should be kept in mind that the one-third of bananas used for beer, also supply calories and nutrition.)

Only recently, have the Haya become large consumers of carbohydrates other than bananas. My food survey shows that cassava was the main course in 20% of surveyed meals, maize in 15%, potatoes or sweet potatoes in 10%, and rice in 5%. The use of these foods signals not only a shift in dietary habits but also a decline in subsistence agriculture as a way of life, since a portion of cash incomes is used to buy maize flour, rice and cassava.

About one-fifth of villagers cash incomes comes from brewing and distilling, therefore beer bananas are an important source of cash used to purchase food. About one in ten meals are financed by beer bananas and related alcoholic beverage sales. Banana beer itself supplements the diet of mainly male drinkers.

Food Shortages

Bananas are propagated from cuttings: after planting it takes one and one-half to two years to produce a large bunch of bananas (Kabwato, 1976, p.51). If managed correctly, bananas are an ideal food crop because they bear fruit throughout the year. Timing the planting, thinning the unpromising suckers and pruning the outer layers of the stem, are the traditional survival skills passed on from father to son. By the 1980s, these skills had declined drastically. Tibaijuka (1979, p.123) reports that three-quarters of her male respondents under thirty-five were not very good at regulating fruiting. Tibaijuka (1984) estimates that ten percent of the banana harvest is marketed: most of these transactions are frictional purchases and sales which cover the temporary shortfalls due to irregularities in fruiting.

Moreover, Districts with normally adequate food security may experience seasonal shortfalls due to poor climate or pests. Beer bananas are more pest resistant so are becoming more popular in the more densely populated areas, where the banana weevil and nematode problem is acute. Decreasing land productivity and crowding has produced an underclass of poorer farmers with inadequate land to supply the family's minimum caloric needs. Tibaijuka's (1979, p.35) estimate of overall regional food production (five kilos of bananas per person per day) may be more or less correct but it gives the false impression of adequate food for all, because the aggregate figures mask shortfalls in some areas and wastage in others.

Tibaijuka provides an excellent illustration, farmers in one Karagwe District village chose to compost 8,217 bunches of bananas rather than sell them below the cost of cow manure (Tibaijuka, 1984, p.123): while families in other areas either purchased bananas or replaced them with substitutes. Ngaiza, claims that as much as forty percent of the entire regional banana harvest may be wasted due to the poor distribution network and the inability to control the exact time that the bunches reach maturity (Ngaiza, 1981, p.103). There is no doubt that inadequate transport and poor regional distribution is a very serious problem. Some wastage could be prevented by leaving the bunches on the trees for a longer time but there is a well founded fear that these bunches will be stolen or destroyed by hail (Ngaiza, 1981, p.85). During the 1994-1996 period after the arrival of a half million Rwandan refugees into the Karagwe and Ngara districts, the demand for plantain was much greater than the supply and these districts had no a wastage problem. The Haya have names for three types of food shortages:

- *olujala,* is a seasonal shortage of bananas, usually in the November to January period,
- *enjala,* is a time of generalized shortage of all foods, which are obtainable but in short supply and very expensive,
- *eifa,* a period of acute food shortage when people are reduced to eating non-foods, such as, banana roots or red sorghum, normally reserved for beer brewing. (Ngaiza, 1981, p.82)

During a fifty year period, villagers in Bushagara, Muleba District, reported twelve years of serious food shortages: 1929, 1930, 1935, 1940, 1946, 1949, 1954, 1960, 1966, 1969, 1974, and 1979. The famines of the

1930s were the most serious (Ngaiza, 1981, p.29), the other times were closer to *enjala*.

Ngaiza reports that in the pre-colonial and colonial periods, the chiefs, maintained food security by collecting bananas during times of surplus and had them dried and placed into storage (in banana sheaf containers) in protected huts. In post colonial Tanzania this traditional form of famine relief has disappeared and there is no current solution to the interrelated problems of banana wastage, overselling when refugee camps create a boom market, lack of storage, poor transport and inadequate distribution. Storage did have its drawbacks, two elderly gentlemen, between seventy and ninety reported that dried bananas tasted terrible and would only be used as a last resort.

Fifteen percent of my 250 household sample are permanently at risk because they have insufficient land (less than two acres) for the household (three or more persons) and a small cash income (less than $500 Tsh equivalent per year). This group has the least food security because they have an extremely small margin of savings and they do not produce either enough food or cash to ensure even their minimum calorific needs.

Peasant Laziness? The Colonial Assumption

Too often, rich farmers and government officials complain that most peasants are lazy and therefore do not bother to properly cultivate either food or cash crops. This complaint echoes the view of an earlier generation of administrators, the colonial officials, who tended to blame food shortages on Haya drunkenness and laziness, and lack of enterprise.

In 1935 C. McMahon the Provincial Commissioner, reported that the coffee trees were in bad shape. In 1936 he stressed that the Haya should improve cultivation to obtain higher yields. One year later, the P.C., G.F. Webster stated:

> ... the banana supplies them with both their staple diet and their beer, the latter of which they drink in excessive quantities ... coffee, which is grown among the bananas, is their cash crop and

> since neither the banana nor the coffee tree require much labour to maintain, the people have acquired a very indolent habit of life.

In 1943 the colonial officials voiced the same complaints although noting some improvement:

> Cultivation of the coffee plots generally were maintained at a reasonable standard, special attention being given to improvement of the soil. A mulching campaign was undertaken ... and results can be said to be satisfactory.
> Care of the coffee tree was not what it might be, but instructions in regeneration and care of the tree continued to be given and progress reported. Today it is extremely rare to see a Haya banana/coffee plot that is not mulched and weed free. If this practice is the result of a colonial extension campaign begun in the 1940s then this is excellent evidence that the Haya *will* adopt innovations of proven usefulness.

In 1947 the PC, R. Z. Hall echoed McMahon's complaint of 10 years earlier:

> The African coffee growers in Bukoba, with a few exceptions, have always expected their crop to look after itself and have been unwilling to undertake recommended methods of coffee culture and pest control.

In 1953, officials repeated the same complaints:

> The Buhaya take little interest in the regeneration of new coffee wood and the control of pests, though attention to these could almost double the yield in a season.

Again recommending that the Haya should pay more attention to tree maintenance and to preventing the depletion of soil fertility. A later FAO report on world coffee production makes similar comments. Krug noted that:

> ... production could be doubled in five years merely by pruning, spraying and better picking. (Krug, 1968)

Since coffee is grown interplanted with bananas, the same arguments are used when referring either to cash crop or food production.

Goran Hyden considers peasant behaviour not as laziness but as adherence to a different value system. Hyden argues that if incomes rise because of high prices or productivity increases, peasants will increase their leisure rather than try to produce more surplus because there is a greater premium placed on social activities and the economy of affection than accumulation of wealth. Hyden's theory is of tremendous explanatory value, but the concept of "peasant laziness" or the "economy of affection" should not be used to explain a situation in which peasants have no economic incentive to produce more food or cash crops.

Kabwato (1976, p.77) provides an alternative cost/ benefit explanation of Haya agricultural practices. The farmer's first three hundred hours of labour on a one acre banana grove will provide a $67 (Tsh equivalent) income, but after this point, the marginal returns to labour decline drastically. If the farmer puts in one thousand labour hours on the same acre, his or her income will only be $75 (Tsh equivalent). If we assume (approximating the actual situation) one labourer per acre, and three hundred working days per year, then the farmer who works one hour per day on the core plot will receive an income nine-tenths as large as the farmer who works ten hours per day. But in fact, the "lazy" farmer working one or two hours per day on the banana shamba can also work at petty trading, handicrafts or other jobs, and earn more income than that from additional work on the small average land parcel. Every one of the interviewed two hundred and fifty heads of household, worked and earned cash outside the core plot (*kibanja).* Two-thirds of all cash incomes came from trading, salaries and artisanal activities. Rald and Rald (1975, p.97) Tibaijuka (1984, p.100) claims that 54% of Haya incomes are derived from alternatives to coffee and bananas. The higher proportion of alternative incomes in my study, can be explained by two factors. A decade later there is a greater integration into the market economy because of loosening restrictions on private trade as part of the IMF's structural adjustment program. Secondly, Tibaijuka surveyed villages outside the region's commercial and population core, however my informants lived in the "lake littoral zone".

The real issue is not peasant laziness, but land shortage and the attractiveness of alternatives to labour on the core plot. As Kabwato points out, the farmer is a target worker who has very little incentive to put in more than about one hour per day per acre. Since few farmers have more than one acre per worker, the rational choice is to limit farm work and seek alternative employment. In fact the rich farmers and officials, who complain about the laziness of their poorer neighbours, usually hire them in order to get their own coffee harvested and keep their plots in good order; since their time is more highly rewarded off the farm. Few women have time for drinking *pombe* and socializing, and even the men who do the most drinking and socializing may not have any more profitable alternative outlets for their time and money.

Beer Bananas, Brewing and Related Activities: Recent Innovations

In the past food security depended exclusively on subsistence agriculture, while today it is tied to many activities, especially the production, consumption, preservation and sale of both food and beer bananas. Beer banana varieties use up scarce resources in the densely populated areas but they also enhance food security because they are more disease resistant and the income from brewing and related activities provides money which can be used to buy food during times of plantain shortages. Even the youngest children derive some nutritional benefit from brewing since they are fed sorghum removed after the beer is decanted from the fermentation vats.

The monetization of the regional economy along with deterioration of the core farming areas has meant a tremendous growth in beer and liquor trading. This in turn has resulted in a series of new production methods and innovations in marketing.

The Evolution of Regional Brewing

Beer brewing probably began soon after the first banana cultivation, or the "time of the ancestors" as one informant put it. As population density grew and as the spread of the banana weevil and related pests decimated the food banana crop, the importance of brewing increased. As an adaptation/survival strategy, many Haya switched to the *kishubi* banana — a hardy, weevil resistant variety originally imported from Uganda (Tibaijuka, 1984, p.117). The large-scale cultivation of *kishubi* bananas dates back to the time of the First World War, when it was adopted by Chief Kalemera of Kianja and Chief Muhiga of Kiziba kingdom. These chiefs were reputed to have introduced the practice of boiling down *orubise* banana beer to make *konyagi* banana liquor. This process increases the alcohol content from 1-3% in *orubise* to 10-15% in *konyagi;* this has the disadvantage of increasing the potential for drunkenness, but the advantage of allowing for longer storage life, less waste of resources, and fewer possibilities of infections.

At first, *konyagi* production was monopolized by the chiefs, who limited consumption to their courts and Hinda clan festivals. They enforced their monopoly with strict penalties. By the time of the Second World War, the Chief's no longer had either the moral authority or the absolute power to impose harsh penalties: *konyagi* use increased, until by the 1950s, there was already widespread private consumption by commoner clans. After Independence in 1961, *konyagi* became a common commodity even though it is officially outlawed.

Coinciding with the spread of *konyagi*, came another important transformation of Haya social customs, the widespread purchase and sale of banana beer. Before World War Two, *orubise* beer was generally shared with guests. The head of household would fill his personal kalabash and then he began the round of drinks. By the late 1960s, *orubise* began to be sold in the household, and by the beginning of the 1980s, the old style of redistribution had virtually stopped. Now it is extremely uncommon for Haya to serve a substantial quantity of beer or liquor without demanding a cash payment. Haya beer drinking seems to be more highly commoditized than elsewhere in tropical Africa, although the scanty literature on the subject makes it a largely unexplored research area — the few studies include Heritier, 1975; Saul, 1981; Wolcott 1974.

In Kagera Region, beer and liquor is now permanently distributed in a variety of venues outside of household hospitality, these include: the sitting rooms of some houses, which have become *ekigata*, informal sector bars; village general stores, which provide locally made beer and sometimes liquor by the glass or bottle; specially constructed kiosks (usually of mud and thatch) on pathways where large numbers of passersby congregate. There are very few social restrictions on the distribution of beer and liquor, to local residents. Because liquor sales are illegal, a young unknown male entering a village drinking venue, probably would not be served, in case he was a police informer, vigilante (*sungu sungu*) or a thief.[1]

Men may stop for an hour or two at a kiosk or duka, after completing their days work in the fields. Women are too busy at this time carrying water, collecting firewood and preparing the evening meal.

The Economics of Brewing and Distilling

In the 1990s the trend continues to commercialization of banana beer and distilled alcohol. In the evenings the Haya men and a few women congregate at *ekigata*, informal sitting room bars, which can accommodate twenty or so people at any given time. The *ekigata* is situated in the front room of typical Haya house: homes are usually rectangular, mud-plastered mangrove pole structures, covered by iron roofsheets. Usually the door, and the front windows are left open. The bar proprietor places a woven banana leaf head protector on a stick, by the nearest pathway, as a sign that beer, or beer and liquor, are being sold.

Gration Kabalulu, one of my research assistants, and myself visited twenty *ekigata*, and took stock of their basic setup. We observed that on average, a home-based bar would contain an inventory of: seven wooden seats, two large straw seating mats, one kerosene lamp, three *karoboi* (small candle-like lamps made from old tins), ten glasses, nine bottles, two washing buckets, three gerry cans, and a radio cassette player. Houses that are not being used as bars contain the same items but fewer of them: the stock listed above was worth one hundred and fifty dollars, based on 1988 prices. This was double the value of equivalent goods at an ordinary household. The average client drank three litres of beer (including the beer content of boiled down *konyagi*) and spent about $1.50 US, in Tanzanian currency. The usual pattern was for clients to drink until either their money or the beer and liquor ran out.

Haya beer drinking is on a par with other regions of Africa, South of the Sahara (FAO, 1975). I estimated per capita beer consumption to be three-quarter litres per day or 273 litres per year. This estimate is based on regional population and banana production statistics and my own participant-observation study of local bars. The average mainly male population of regular drinkers consume about three litres per day of beer (or one-half litre of *konyagi*, boiled down beer). However women consume beer in much smaller quantities than men, and children hardly at all. There is also a small minority of male non-drinkers such as devout Muslims, Seven Day Adventists and evangelical Protestants who lower these average consumption figures. This estimate and others like it (for example 236 litres of beer per year in Burkina Faso, (Saul, 1981, p.747) are merely parametric figures distorted to some degree by sampling inaccuracies. However my regional estimate corresponds to the servers tally in local bars in which the average was in fact three litres of beer per customer.

Beer, is brewed in a boat-like, three hundred litre wooden vat called an *obwato*. Brewing one boat requires an average of fifty hours of labour (Rald and Rald, 1975, p.97; Tibaijuka, 1984, p.258) and US$ 18 in costs, for materials (bananas, sorghum and firewood) and rental of equipment.

The work of brewing involves: cutting the bananas and carrying them to the boat, ripening the bananas, roasting and malting the fermentation agent *ulezi* (red sorghum), fetching two hundred or so litres of water, mashing the bananas by treading on them, supervising the fermentation process, cleaning the utensils and bottling and selling the beer.

Cutting and carrying the banana bunches is traditionally considered the job of the male head of household, but because of pressure to make large quantities of beer in the least time, women often do some or all the carrying. Out of fifteen local bar owner/brewers: one woman bar-owner did all the carrying herself; in four cases wives and children did some of the porterage; two others used hired labour; and one brewer carried bananas by tying them on his bicycle.

Ripening the bananas can be done in two ways. The traditional method is to tie the bunches to a pole placed above the cooking fire for a week. Larger brewers use a newer method which only takes three days: the bananas are placed in a hole in the ground, covered over, and a low fire is burnt on top. They can ripen as many as sixty bunches of beer bananas at one time and therefore brew about two hundred litres of beer per day for processing into liquor.

The Haya consider treading on the bananas and supervising the fermentation a skilled job which can only be performed by males. Female brewers pay a skilled male about two dollars (double the minimum wage) for this one day operation. The remaining tasks (water fetching, roasting sorghum, etc) are carried out exclusively by women, giving credence to the observation that even though brewing is considered a male job, most of the actual work is done by women. (Swantz, 1985, p.56)

The variable costs in the brewing process are:

- beer bananas, TSh equivalent of eight dollars;
- firewood, underground ripening method, one dollar;
- fifteen kilos of sorghum, six dollars;
- hire of the *obwato* or "boat" most often paid in kind with 10 litres of beer or it can be purchased for forty dollars;
- depreciation on the equipment, which according to the calculations of Tibaijuka (1985, p.258) amounts to another $1.50.

If labour costs are calculated at 10 cents US per hour (the minimum wage scale) then the total variable cost of one brewing session is five dollars. This makes a hypothetical total cost of production of twenty three dollars for one boat (250 litres) of *orubise*. This cost is hypothetical because the small farmer and family, usually grow the bananas and do all the work themselves, rather than lay out scarce cash.

Farmers do not calculate family labour costs at minimum wage rates because they do not have alternative wage options. The average farmer/brewer grows two-thirds of the bananas, gathers the firewood from communal areas, considers depreciation on brewing vats as normal farm expenses, and does not take into account the unpaid labour of women and children. They calculate the "costs of production" as less than ten dollars, that is, the price of the sorghum, use of the *obwato;* their own labour is not considered a cost if it has no alternative economic value.

One *obwato* of beer can be converted into ten 25 litre lots and sold customers who pick it up in their own gerry cans: this is the equivalent of wholesale marketing and in 1988 would yield the producer twenty five dollars revenue ($2.50 per 25 litre gerry can). Using the "smallholder calculation" brewing beer yields an income double the costs and is a "profitable" way to mobilize family resources.

About one-quarter of the retail *orubise* and *konyagi* sellers do not brew beer themselves, so that they can specialize in liquor making and retail sales. Retail prices are ten cents US for a one-half litre bottle of beer, therefore the sellers take in double the purchase price, five dollars (TSH equivalent) for each gerry can they sell. The bottling, washing and serving is usually undertaken entirely by unpaid family members. Taking into account occasional uncompensated loss of bottles and glasses, the sellers reckoned that their time is fairly compensated.

Many beer producers boil down their *orubise* to make *konyagi*. This can be done in two ways. For small scale production, the apparatus is made from two large round pots *(sufuria)* joined together with wire, using a gasket of dried banana stem fibre and an inserted bamboo outlet pipe leading to a cooling basin. The advantage of this method is that it is cheap; it does not require either expensive equipment or a large input of beer or firewood. The disadvantages of this method, are that only small lots of three or four litres can be produced at each distillation, and it can be dangerous. Informants reported that accidents sometimes occur when the pots separate and boiling beer shoots out.

When larger quantities are produced, the preferred method — a recent innovation — is to boil the beer in a two hundred and fifty litre metal drum. The distillate flows into a large bamboo pipe sealed at the end, and is collected through ten or so outlet bamboo tubes which flow into clay pots, set up in a stream or flooded ditch for quick cooling. Using this method, the contents of an entire *obwato* can be rapidly processed; the two hundred and fifty or so litres of beer yield seventy-five or eighty litres of liquor, which is diluted with five to fifteen litres of water to increase the volume of the final product.

The costs of processing ten gerry cans of beer are about thirty-four dollars:

- drum rental $3 (the drum costs $US 30-60 to buy),
- eight bunches of firewood $4,
- five hours of supervision $1,
- depreciation on clay pots and bamboo $1,
- ten gerry cans of banana beer $25.

The ten gerry cans of beer yield three of *konyagi*, which if sold in bulk, will fetch about forty dollars: this is a six dollar profit. But most *konyagi* producers don't purchase firewood, labour and banana beer, so that they don't include these as costs in determining revenue.

If liquor is sold in three-quarter litre bottles, at fifty cents US per bottle, the seller will take in fifty to sixty dollars. Obviously, the more integrated producer, using "free family labour" to grow most of the bananas, to cut their own firewood and to carry out the other tasks, can realize the highest rate of return. Those who buy banana beer, as an input in liquor production, can specialize in retail sales: their lower rate of profit is compensated by a larger overall amount of sales and profit.

The Pombe Village

An assistant and I studied informal bars in a "pombe village" — a site where specialized brewing and distilling takes place on a large scale and alcoholic beverages are readily available. Village elevation is at Lake Victoria level, 1100 metres, this allows cultivation of red sorghum which is difficult to grow on the higher ridgetops — the scarcity of sorghum is often a limiting factor in beer production (Ngaiza, 1981, p.73). The area is plagued by an infestation of banana weevils, less common at higher elevation. This in turn, ensures that more farmers produce beer bananas, as a cash crop.

A weekly farmers market offers a plentiful supply of sorghum and beer bananas for sale, as well as bringing together farmers with money in their pockets and things to discuss over beer and liquor. The village is at the base of a cliff where springs and streams are plentiful, assuring water for beer and pools for cooling drum-made *konyagi*. The stream beds, not usable for agriculture, supply an adequate supply of sieve grass (used in brewing) and firewood, both considered communal property. The stream beds also chop up the terrain making it difficult or impossible to reach by police motorcycles or Landrovers. Finally, the village is only five kilometres from the main Bukoba to Mwanza trunk road, this increases passing trade and advantages for quantity selling. Brewing and distilling go on everywhere in the region, but in the "pombe village" production is carried out on a larger scale.

In the pombe village, there is one officially licensed bar that sells Tanzania breweries bottled beer. Because of limited supplies and high prices, this bar is infrequently open for business, whereas on any given night twenty to thirty informal sector bars will be filled to capacity.

The typical bar owner (based on a sub-study of twenty bars in the pombe village) is: a) male (16/20); b) married (15/20); c) Roman Catholic (17/20); d) lists his primary occupation as farmer (19/20); and e) has completed a median of five years of primary school.

The twenty surveyed *ekigata* could be demarcated into three groups based on sales levels. The six bars with the lowest sales earned a net annual income of between $350-599 US from their operation. The middle seven, earned between $600-899: and the most successful seven earned $US 900-1700. The primary distinguishing characteristic of the richer group is their larger volume of total retail liquor sales. All the high income group produce their own *konyagi*, and all own an *obwato* and a drum — this gives them a relatively large productive capacity. In contrast, only one-half the poor and medium scale owners produce *konyagi,* and, only one-third of them have access to a brewing boat and a drum.

I was surprised that there were no significant differences in seating capacity, products sold, entertainment, etc., between the more and less profitable bars.

Conclusion

The beer and liquor trade, demonstrates that when it is in their interest to do so, and when the scale is appropriate, small farmers will readily adopt innovations. Village brewers quickly changed their traditional methods in order to increase their profits by using: a new banana variety, new ripening methods, new distilling techniques, and a larger scale of retail sale. The beer and liquor trade, is a good example of local grassroots economic initiative and technical innovation (such as the use of metal drums in distilling), but it is also a source of alcoholism and related social problems see (WHO, 1980; Room, 1984).

Production and trade in alcoholic beverages is intimately related to the issue of food security. Dense population in the most desirable districts, facilitates the spread of nematodes and weevils, moreover it intensifies land degradation which leads to declining production of the traditional food bananas. Farmers shift production to hardier beer bananas because they are pest resistant and because they can make a living by selling beer bananas, beer, liquor, or any combination of the above.

Project planners can certainly learn something from the success of the *pombe* trade. The reasons for its success include the following:

- the small requisite start-up capital, allows disadvantaged farmers and women with little land to earn a living;
- the skills needed to produce beer and liquor are known by most farmers;
- social drinking provides an outlet for information exchange and recreation, consistent with traditional culture;[2]
- costs are low relative to mass produced alcoholic drinks;
- the emphasis on exploiting cheap local raw materials;
- the use of simple technology;
- decentralized production by family units;
- response to a local market;
- involvement of both sexes and different age groups;
- distribution mainly in a local area with no need for motorized

transport.

The disadvantages of the enterprise are primarily associated with alcoholism:

- increasing alcohol consumption, reduces savings and accumulation;
- impoverishment and indebtedness for those with the most serious alcoholism-related problems;
- social problems (prostitution, wife and child abuse, family separation);
- lost labour productivity;
- alcohol related diseases, the spread of AIDS and other venereal diseases, increased health care costs, and so on.

In some ways the industry and innovation shown by Haya beer and liquor producers belies the notion of peasant laziness and traditionalism. Similarly the strict cash sale practices of the village bars do not support the notion that the economy of affection is the primary motivator for economic decision making. Yet the trend toward increasing production of stronger brew may also mean a trend toward long-term economic decline as a result of increasing alcoholism and alcohol-related problems.

The banana liquor cottage industry contains within itself health and social costs, moreover the "economics" of investment and savings are unclear in a trade where, sometimes, the largest producers and sellers are also the largest consumers.

The roots of the illegal liquor trade and its accompanying problems are to be found in the declining food banana production and the dearth of alternative opportunities for the small and middle farmer to earn cash needed to build up the farm and to *buy* food supplements. Regional beer and liquor production is likely to increase until significant progress is made in solving the problems of land shortages, environmental degradation from overgrazing, banana and coffee pests, declining food production, and, the lack of alternative cash generating opportunities.

The only possible way to reduce the production and consumption of alcoholic beverages is the promotion of substitute economic activities with an equal or better rate of return to labour and capital. Such initiatives must satisfy local consumer preferences. Small farmers need loan capital and training in marketable skills such as tailoring, furniture making, construction, catering and others.

Notes

1. The information in the above two paragraphs is based mainly on personal communications from N. J. Mutimbo, A. T. Lugambwa, E. Bagambi, J. Bubewa and G. Kabalulu.

2. Informal sector bars serve this same function even among Africa's most proletarianized social strata, migrant workers in Southern Africa, see for example (Crush, 1988).

8 Coffee, the World Market Connection

The Haya farm is an integrated enterprise. A typical household, cultivates food crops (bananas and beans) cash crops (mainly coffee and beer bananas) and also supports itself with income from extra-agricultural enterprises, such as fishing, house construction, furniture-making, tailoring, watch repair, brewing, distilling, village bars, petty trading, traditional medicine. Village traders and artisans supply all the locally available goods and services. Family visits into and out of the villages provide an injection of commodities either not available locally or too expensive for everyday use.

To outsiders, including some researchers, coffee growing appears to be the primary source of cash incomes because it is the only export and therefore the only large source of cash from outside the region. In spite of what some of the literature and the official statistics indicate, I found that coffee revenues never exceeded a third of total cash, even in villages in the heart of the coffee belt. The coffee purchasing agent is the KCU, the Kagera Cooperative Union: the link between the smallholder and the world economy. Although the KCU and the government of Tanzania are not always in agreement and although they *compete* for coffee revenues, nevertheless they both *depend* on coffee revenue. However, Haya farmers categorize coffee as one among many sometimes competing sources of cash.

The Tanzanian Government's master plan for the coffee industry is to double its production bringing it up to about one hundred thousand tons by the year 2000. To support its objectives the government launched a number of coffee related projects in the 1980s, assisted by financing from the EEC. The Coffee Development Programme aimed to: provide eight million certified coffee seedlings annually; to subsidize inputs — sprayers, insecticides, coffee tray wires; to support extension services and education; to rehabilitate coffee related transport; to increase measures to prevent coffee berry disease, leafrust, and other diseases (*Daily News*, July 10, 1988). Since the early colonial period successive governments have considered coffee growing as a way to generate much-needed revenues and foreign exchange. Yet Haya farmers tend to disregard or reject the "grow more coffee" campaigns of successive governments. The reasons are rooted in the history of the region, and in many ways the present campaign echoes previous failed attempts by the British Colonial Administration to control peasant coffee production. The present productive capacity was established in the late 1920s by smallholders who decided how much coffee to grow and how much time and effort to devote to coffee trees.

The Nation's Number One Export

Coffee is Tanzania's most valuable export crop. Between 1980 and 1987, Tanzania exported an annual average of 52 metric tonnes of coffee, which provided the foreign exchange starved nation with roughly 35% of the total value of all exports (ERB, 1988,p.50). The bulk of Tanzanian coffee is produced in three regions, Kilimanjaro/Arusha, Kagera, and the Southern highlands. Kilimanjaro/Arusha is the only area with a substantial estate sector: there has been an absolute and relative decline in the estate sector over the past two decades. Slumping plantation output and a doubling of smallholder acreage has reduced the percentage of output from 24% to 5%, *Daily News,* Dar es Salaam July 10, 1988. Plantation coffee accounts for only five percent of national output. Kagera region, supplies approximately forty percent by bulk of the nation's coffee or fourteen percent of foreign exchange earnings at the time of the 1988 census.

Simon Mbilinyi, highlights the major problem of Tanzania's coffee exports; too many resources are concentrated in one industry where the primary determinants of supply and demand are beyond local (and national) control. World coffee prices reflect many determinants: global weather, the four year time lag in responding to the market, the natural cycle of tree crops, and intense competition among fifty producer countries in the South, for a market in the industrialized North (Tanzania controls less than two percent of world output).

Despite these problems, Simon Mbilinyi's prediction that "the coffee industry will remain dominant in the near future" (Mbilinyi, 1976, p.205) is still true more than twenty years after his fieldwork, and for the same reasons:

> . . . the price of coffee is relatively higher than most ... alternative crops for diversification, even when world market prices for coffee are low. (Mbilinyi, 1976, p.203)

The Tanzanian Government's plan for the coffee industry — to double total production to 100,000 metric tons by the year 2000— is clear evidence of coffee's ongoing attractiveness.

For forty years, the seventy-four member nations of the ICO, the International Coffee Organization, tried and failed to stabilize supply by instituting a quota system. Since the ICO could never enforce quotas it finally decided, in October 1989, to abandon this system until October 1991, and subsequently the ICO dissolved. Since then, about two dozen African and Latin American countries formed the Association of Coffee Producing Countries which planed to reinstitute a coffee cartel arrangement, but has failed to implement an effective system. Frost in Brazil in 1994 raised coffee prices to an all time high and in the scramble to cash in, coffee producers were not willing to cooperate. Even when the ICO regulations were supposedly in effect, Tanzania did not adhere strictly to its 50,000 tonne quota; the government set its 100,000 tonne coffee objectives even before the 1989 world coffee crisis.

In 1989, the average coffee price was $1.25 US per pound, but in July, the price dropped to a fourteen year low, reaching 70 cents US per pound. The reason for this slump was speculation (proving correct) that the ICO would not ratify its quota arrangements. This price was below the average cost of production for African countries, such as Ivory Coast, which rely on estate sector coffee, but above the opportunity costs of Tanzanian peasant-grown coffee.

In its endeavours to bolster coffee production, the Tanzanian Government must solve three interrelated problems: 1) to plant enough trees so as to establish the desired productive capacity; 2) to produce the desired amount of coffee, meeting industry standards of FAQ (fair average quality); 3) to try to stabilize national, and local incomes in the face of fluctuating short and long term world market supply and prices. Only Brazil, can unilaterally change world coffee prices because it controls about one-third of the world's supply. Supply of coffee, is not simply a matter of decisions made by national leaders. It depends on: producer prices, marketing arrangements, farm management, inputs, the weather, smuggling, and a host of other complicated interconnected factors.

The History of Coffee in Kagera Region

Some coffee was grown in the region before colonization, but the extent and methods of its cultivation changed dramatically after colonial rule was imposed. Prior to the 1890s, peasants picked green berries and prepared them as coffee balls (Jarvis, 1938): after 1890, coffee was harvested for export. From 1891, the inception of German colonial rule, through the British period, the Haya peasantry became important commercial coffee producers — they produced 60% of Tanganyika's coffee by volume prior to Independence, see table 1. The colonial state, missionaries, and the few white settlers in the region encouraged smallholders to grow coffee, the 1920s coffee boom was the direct result of contact with these outside agencies. At the end of the "coffee boom" of the 1920s the Kagera Region was already producing as much as in the 1990s, twelve thousand tons of coffee.

1922......2,900	1944......6,100	1967......13,900
1923......2,600	1945........8,100	1968......12,000
1924......3,600	1946........4,400	1969......14,600
1925......4,100	1947........9,500	1970.......9,200
1926......4,700	1948........5,800	1971......14,000
1927......3,900	1949........7,500	1972......10,800
1928......7,800	1950.......11,500	1973......12,800
1929......6,800	1951.......11,700	1974......17,000
1930......7,400	1952........9,300	1975......13,200
1931......6,600	1953.......11,000	1976......15,300
1932......7,100	1954.......10,200	1977......13,200
1933......7,900	1957........9,000	1978......13,900
1934.....10,200	1958........9,200	1979......13,700
1935.....10,900	1959........8,100	1980......15,200
1936......6,500	1960........8,400	1981......16,000
1937......9,500	1961........5,300	1982......17,600
1938......8,300	1962........8,900	1983......14,900
1939.....12,000	1963........5,400	1984......10,700
1940......7,800	1964........8,300	1985......11,300
1941......6,700	1965.......14,300	1986......12,900
1942......8,100	1966.......10,500	1987......12,300
1943......7,300		

Table 1: Kagera Region Coffee Exports
(all varieties, to nearest metric ton)
Sources: Tanganyika Provincial Commissioners, 1922-1954; Kagera Cooperative Union, Statistics made available to author.

The British realized that in densely populated Northwest Tanganyika — far from the Nairobi/Mombasa coffee auctions — there was little potential to create a coffee-plantation economy, run by settlers. By the 1950s, the "Lake Province" (which included the Sukuma lands near Mwanza) contained only 117 European and Asian estate holdings, controlling 18,000 hectares: this figure is in marked contrast to the 183,000 hectares of alienated land in the Kilimanjaro region and 254,000 hectares in the Tanga region. Rather than promoting an impractical settlement policy, the British, opted for a strategy of maximizing smallholder coffee production.

At first, only the most privileged farmers planted coffee. In the pre-colonial Haya kingdoms, the distribution of land was based on a complex system of customary law. Haya society was stratified into two castes: a higher *Hinda* royal/landlord class, and a lower *Iru* commoner/peasant class. The Hinda caste controlled large plots of land, from which they extracted labour and produce rent. During the colonial period, a class of richer peasants accumulated larger amounts of land — some alienated by the colonialists, and some acquired under traditional feudal tenures. When a market for coffee developed, chiefs, and larger landholders, opened all available land, planting coffee "wherever it would grow". During the era of coffee plantations:

> ... land which up till that time had not been cultivated, was planted with coffee wherever it would grow. ... about 50 years ago when coffee became recognized as an economic crop, the chiefs took over large areas of land capable of coffee bearing, or deserted banana plantations, and put them under coffee. The work of preparing and planting these areas was done as *nsika* (forced labour) by the chief's subjects. (Cory and Hartnoll, 1971, pp.279 and 149)

By the 1920s, high coffee prices and relatively cheap and available consumer goods enticed most Haya farmers to grow coffee. During the "coffee boom" they planted millions of trees in Kagera Region. The British administration never actually counted the number of coffee trees, therefore, estimates varied wildly. In 1939, the Senior

Agricultural Officer, T.S. Jarvis, claimed that the Bukoba district contained three and one half million trees (Jarvis, 1939, TNA): in 1932, Armitage-Smith reported a tree population of two million. Based on coffee export figures, both appear to be underestimates. By 1928, Haya exports reached about 8,000 tons, which means that there were likely between twelve and sixteen million trees. Friedrich (1968, p.161) measured an average annual production of 0.7 kilos of peeled coffee per tree (in his study of 120 farms) and, official production statistics today report a average per tree of just 0.5 kilos (Raikes, 1976,p.6). A mature coffee tree may produce about two kilos of hulled coffee cherry per year: however Bukoba production did not take place under optimum conditions and there was quite a high level of rejection of poor quality coffee cherry.

The 1955 annual report of the Tanganyika, Provincial Commissioners claimed that, there were 92,000 growers. In 1954 exports were just over 10,000 tons, and if the reported number of coffee growers was correct, then the average Haya farm was growing just over one hundred kilos of coffee. Friedrich (1968, p. 191) reported an average tree density of 93 per acre, and as noted, a harvest of 0.7 kg. per tree, if these figures are accurate, the average Haya coffee grower of the 1950s kept about 150 coffee trees on an average plot — under two acres of interplanted land.

During the period of British colonial rule the administration tried to enforce standards of coffee production which, had they been carried out, would have threatened the interrelated subsistence/cash crop equilibrium. These directives, the "coffee rules" — meant to regulate production on the small hold farm — provoked farmer's protests and were never implemented. Haya resistance culminated in the coffee riots of 1937. After this date, the colonial state came to realize the futility of direct intervention on Haya farmsteads; neither the British nor the post independence state ever attempted to *enforce* farming methods or standards.

The 1930s attempts to force smallholder coffee growers to follow specific agricultural directives led to direct resistance and violent confrontation: the cause of the confrontations were "the coffee rules" — which posed a direct challenge to the diversified peasant farm enterprise.

The regulations were part of a coffee marketing and distribution system in which the small farmer felt powerless and exploited. Under this system, more than five hundred "itinerant traders" (the vast majority of them "Arab" or "Asian") bought the crop from the small farmers and then shipped it to Mombasa for auction purchase by the agents of British firms — Twinings was the largest (Leubuscher, 1939). Most farmers felt exploited because of irregular pricing policies and the resulting large price fluctuations; producer prices were kept as low as possible and did not correspond exactly with the ups and downs of the world market. In 1982, I interviewed twenty coffee farmers of the 1940s and recorded the following comments, which are typical:

...price was set by the Asian businessman depending on supply, so it fluctuated from morning to evening...
The prices were fluctuating within hours, depending on the number of sellers who went to sell their coffee.
The system was not fair, as the price fluctuated within a matter of hours.
We felt exploitation as prices did change from hour to hour within a day.
We felt the exploitation as they paid the prices they wanted without considering the work done.
The price was set by merchants and fluctuated from time to time; it even varied from morning to evening.
The price was set by the merchant and it varied from time to time, even from merchant to merchant.
We were lowly paid and prices fluctuated from one day to another.
Prices were fixed by the Asians; they fluctuated from one hour to another, depending on supply.
The price was set by the merchant, each according to his wish; it was not constant.
Yes, we felt the exploitation as prices fluctuated within hours.
We felt the exploitation as prices kept changing day after day.

The inherent problem with this system was that neither the farmer, who felt exploited, nor the small merchant, had any incentive to implement quality-control measures to achieve industry standards (FAQ) fair average quality. By the 1920s, traders at the main lake port of

Bukoba would "forward contract" with Mombasa exporters to deliver set shipments at specified dates. The administration had no way of controlling either farmer or trader. In order to meet deadlines: under-ripe berries were picked, the coffee was inadequately dried, it was packed carelessly along with quantities of stones and dirt, or the bags were stored for months in damp locations and hence the coffee fermented (Jarvis, 1937). The middleman took no responsibility for the contents of the bags, and, the Mombasa exporters usually bought coffee in bond and shipped it. Because of the ready market for Tanganyikan coffee in the early 1920s, almost all production was readily absorbed in London, most of it for resale in southern Europe. But by the late 1920s, standards had deteriorated to such an extent that overseas buyers refused many shipments and thousands of bags had to be destroyed. The quality problem was exacerbated by a world market price drop, and by 1930 Bukoba coffee revenues dropped to one quarter of the 1928 figure (Jarvis, 1938).

The administration tried to legislate a solution to the problem by passing the 1932 Bukoba Coffee Export Regulations and reviving the 1926 Plant Pest and Disease Regulations, amended in 1936. The administration's "Export Regulations" were enforced by random inspection of export coffee to ascertain whether it met international FAQ standards. Many merchants, afraid of having their goods seized, were more careful about the selection of their export consignments. The "Coffee Rules" (the Plant Pest and Disease Regulations), however, were counterproductive and unenforceable.

The rules promulgated in 1926 stated that:

1) trees must be weeded and the weeds burned;
2) bananas must be thinned from coffee plots;
3) diseased or weakened trees must be pulled out and destroyed;
4) coffee trees must be spaced at certain distances;
5) all new coffee plantings must be approved by the chiefs;
6) any sick trees must be reported to the chiefs;
7) no grower must possess picked unripe coffee;
8) old coffee beans must be burned;
9) no drying must be done on the ground;

10) the unhusked coffee must be adequately dried;
11) hulled coffee must be checked for extraneous matter and defective beans;
12) dirty plantations must be cleaned.

Violations of rules 7-12 carried stiff penalties — cash fines for first and second offenses and imprisonment for further infractions.

The 1936 "rules" were even more stringent, stipulating:

1) Only such coffee seedlings may be planted and only such seed planted in nurseries as have first been approved by the Agricultural Department.
2) Every individual wishing to plant coffee must plant not less than 250 trees. All trees should be planted nine feet apart.
3) Before planting up a plot every individual must satisfy the Agricultural Department as to the suitability of the soil.
4) The land must be prepared to the satisfaction of the District Agricultural Officer or his nominee before an individual may plant coffee.

Penalties — Any individual who does not comply with these regulations is liable to have all plants uprooted by the District Agricultural officer or his nominee (Tanzanian National Archives, 1926 and 1936).

These rules were a direct challenge to the basic Haya land-use pattern. The greatest fear of the farmers was that either their coffee trees (which they considered farm capital) would be arbitrarily destroyed, or, they would be forced to extend coffee cultivation beyond the limits compatible with their diversified farming enterprise.

As the letter below demonstrates, farmers wanted to grow coffee, as a way of earning 'profit', but, the "Native Coffee Growers of Bukoba District" who wrote to District Commissioner, MacMichael (on 4 March 1937) were opposed to interference by the Administration: they preferred to allocate farm resources in a way that would maximize benefits within the Haya farming system.

> We do not plant coffee for fire but to get profit, and burning is not the cure of coffee plant disease... We did not grow coffee to become afterwards as a trap to fall in but we grow it for the sake of getting profit and become rich and prosperous... We do not need to be taught how to grow coffee or banana trees or to stop from growing anything on our shambas or on our soil. (TNA, 215/1445, pp.1, 10, 60)

When the British attempted to apply the "rules" in January and February 1937, smallholders would not allow inspectors on their lands, and even took up arms to keep them off: their leaders organized demonstrations in the four largest coffee-growing areas (Bukoba, Ihangira, Kianja, and Kiziba). In January, the Native Administration held a special meeting at its headquarters in Rwamishenye to explain the rules to African cultivators. The chiefs were shouted down, and Francis X. Lwamgira, the single most prominent Haya colonial official, was pelted with stones. Eight people were arrested in the disturbances, their trial several weeks later was broken up by hundreds of armed men. In February, the administration tried to forcibly enter coffee plots in Bukoba district: Eustace Bagwarwa, from the Kianja African Association, organized a armed resistance (with spears) to keep out colonial agricultural officials. The resisters were dispersed by the police and the leaders arrested, but police intervention did not solve the problems or make them go away.

In another incident, Chief Kalemara, of Kianja district was jeered by dissidents when he tried to address an assembly in Kamachumu (the second largest commercial centre in the region). This hostile crowd was dispersed by a police baton charge, and more leaders were arrested. In all, thirty "ringleaders" were convicted and sentenced to short prison terms. After February 1937, the agitation died down and the Kianja African Association went underground (Mwanza Provincial Minute Papers, reported in Austen, 1968, pp.224-227, and Iliffe, 1979 pp.274-286).

The Growth of the Cooperative Movement: Before and After Independence

The coffee riots precipitated some profound changes in the region. During the years of World War Two, the British administration took over coffee marketing from the Asian merchants. By the terms of The Native Producer's Control and Marketing Bill of 1938, the state consigned its agents to buy the entire coffee crop at a set price — slightly higher than that offered by the Asian traders (Bowles, 1979). But the main problem, how to supply a larger quantity of FAQ coffee, was not yet solved. The state decided to encourage local cooperatives, which Haya farmers had begun to organize by the 1920s. After the war crisis period, the state gave up its procuration monopoly; it allowed the private merchants to operate but at the same time encouraged the cooperatives. In 1947, the Bukoba Native Coffee Board was established with a mandate to buy all the coffee from the existing zonal agents. The 1947 Provincial Commissioner's Annual Report pp. 53-54 clearly stated the reason for setting up a local purchasing board,

> The reason for creating the board was twofold, firstly to raise the quality of the exported crop, in which the percentage of rejections has always been very high, and secondly, to go some way toward meeting a strong local demand that the control of the crop should be in the hands of the producer.

In 1949 the Bukoba Native Cooperative Union (BNCU) was founded to centralize the administration of 49 earlier primary societies and associations of farmers established in the 1940s. It took over the functions of the Coffee Board, and by 1951, it acted as a holding company for 67 cooperatives in Bukoba Region and another 7 on the Karagwe plateau to the West. In 1953 the union was renamed Bukoba Cooperative Union, BCU, and its marketing functions were expanded to increase its direct purchases. The administration encouraged the BCU, despite some alleged corruption, because if the cooperatives worked well they could promote progressive farming, through credit schemes, education and extension. In theory, cooperatives were meant to eliminate excess profits collected by the Asian merchants and to redistribute these in the form of higher real producer prices, which would encourage greater enterprise by the farmers.

The organization of the BCU remained more or less the same after Independence in 1961. By 1975, the primary societies affiliated together, took additional marketing powers, and renamed themselves the West Lake Regional Cooperative Union (Kabwato, 1976, p.58 and Tibaijuka, 1979, p.45). Between 1976 and 1982 cooperatives were abolished and their functions turned over to regional marketing boards. This experiment proved a costly mistake, and in 1982 the union was reestablished and renamed the Kagera Cooperative Union (KCU). By 1990, the KCU was the exclusive regional purchasing agent for 203 affiliated primary societies. Under the new government elected in 1995 the coffee industry has been deregulated but not completely privatized so that the KCU must compete with private entrepreneurs who now can buy and process coffee. It remains to be seen whether the KCU will survive private competition and will be forced to completely privatize as has been the case with other parastatals in Tanzania and Kenya.

When KCU was reinstituted, it implemented two major innovations: 1) an immediate and realistic first payment to the farmer, and 2) a policy of only accepting coffee deliveries as "maganda" or "buni" — that is, dried coffee cherries, and not hulled coffee beans. There are several valid criticisms which can be made of the policy of buying only unprocessed coffee.

Raikes (1976) summarizes these: the most important advantage of buying hulled coffee is that it takes up only one-half as much volume when packed, and therefore, the Cooperative Union could reduce its transport costs by a half. The other benefits are: that the farmer could realize the value added income from on-farm processing, and, the coffee hulls when composted, make a good fertilizer and soil amendment.

Nevertheless, in the existing situation the costs of on-farm hulling outweigh the benefits. In 1951, the cooperative purchased BUKOP, a privately owned coffee processing factory, from Asian interests. Since that time, the capacity of the factory has been expanded, to 20,000 tonnes i.e. more than the entire regional coffee crop (Tibaijuka, 1979 p.48). If farmers were to hull their own coffee cherries then the factory would be operating at only a small fraction of its capacity.

From the farmers point of view on-farm hulling is not likely to be profitable because the price differential between hulled and unhulled coffee was relatively small. Mbilinyi (1976) and Raikes (1976) point out that in the early 1970s, when Kagera coffee farmers were receiving ten percent higher prices for hulled coffee, three quarters of them chose not to bother hulling because they could utilize their labour time more productively in other enterprises.

If the small farmer took over the hulling process the cooperative union would loose revenue. First, it would have to pay the fixed costs of the BUKOP factory, which would operate far below its capacity. Second, since it is not possible to supervise thousands of small producers, a significant quantity of coffee would have to be rejected because of poor quality beans, broken beans, stones and dirt in the bags. Third, the coop would have to pay additional costs for extra inspection, grading and quality control to try to standardize the sacks of farm-hulled coffee. The use of composted coffee hulls would not raise overall farm productivity significantly because of the small amounts of compost produced. In short, the KCU would most likely realize a net loss if the farmer were again encouraged to hull coffee on the farm.

Along with its factory hulling policy, the Kagera Cooperative Union has since the mid 1980s been promoting a "grow more coffee" campaign, based on nationally organized subsidies and a policy of higher producer prices. More coffee means more government revenue, so that the aims of the campaign are similar to the British "plant more crops" campaign of the late 1930s (Leubuscher, Colonial Office Library, No. 5713). However, the KCU is unlikely to succeed, since structural problems have in fact intensified in the years since the coffee riots.

There are four fundamental structural components of the Haya farming system likely to constrain coffee production: 1) the demographic structures, 2) smallholder use of farm capital, 3) the emergence of a class system, 4) the marketing/consumption system.

Coffee Production in the 1990s, Demographic Constraints

As we have seen earlier, the demography of the region works against increasing coffee production. The population resident on the farms tends to be too old or too young for hoe agriculture, since only about one-third are between twenty and forty-five. Over one-quarter of households are headed by women who show little inclination to increase coffee production since this is traditionally a male preserve and there are limited incentives offered to them to change this tradition. In the five sub-samples, coffee production also depended on proximity to alternative sources of income: the larger the number of alternatives, the smaller the proportion of coffee income. In the Bukoba town area, coffee production per hectare was lowest. Total coffee production per subsample (expressed in mean kilos per household and percentage of all cash income) was as follows:

- the ridgetop village, 464kg. & 30%;
- the ujamaa village, 470kg. & 29%;
- the coastal village, 505kg. & 28%;
- urban out migrants from the ridgetop, 222kg & 9%;
- the urban satellite village, 155kg. & 6%.

It is clear that coffee production, as a percentage of income, decreases when there are viable alternatives — as is the case with the sub-groups proximate to Bukoba town and the rich farmers discussed in a later chapter.

In the ridgetop village, where coffee production is highest, gender clearly operates as a negative influence. Female-headed households have little space on their land to grow coffee and are faced with disincentives to maintaining even their existing capacity. Thirty percent of the sampled households are female-headed (unmarried, divorced or widowed women who do not have an adult male in residence). The median female-headed farm measures only 0.75 hectares, with the largest recorded as five hectares and the smallest as one half hectare: the average (mean) amount of arable land is one tenth of a hectare (per person) for female-headed households and one half hectare for male-headed households. Only 19 percent of female households replace worn out coffee trees, as compared to 44 percent for those headed by males.

Besides the land-shortage factor, female-headed farms are disfavoured by the customary division of labour, in which planting and pruning coffee trees is a male prerogative. Even in female-headed households the bulk of the coffee-related work is done by males; for example, only 20 percent planted their own coffee seedlings, and only 40 percent pruned their own trees. Only 37 percent of female-headed households, either pruned coffee and/or planted new trees — tasks carried out by occasional hired labour or by male relatives. Only one women in the sample utilized the services of the agricultural extension officer; she paid him to prune her coffee trees.

Of the 213 landholders in the five sub-samples the seventy six female headed households produced an average of 137 kilos of maganda coffee, whereas, the one hundred and thirty seven male-headed households, produced an average of 378 kilos. These figures are very similar to those reported by Bader, for a village in the Kanyigo district; the median male headed household produced 450 kilos per year and the female median was 138 kilos (Bader, 1975, p.150).

The major disincentive to female coffee production was the fact that in more than half the cases the woman did not own her land outright, but held it in trusteeship from the clan. Anna Tibaijuka (1979, p.22) discusses ten legal cases in which deceased Haya husbands had tried to transfer land to their wives. In six of these the ruling went against the widows who were deprived of their land. Similarly, few divorced women have any real security of land tenure. Moreover, in some divorce cases the ex-husband considered the coffee crop his property even though he did not work on the plantation; he might decide to organize the coffee harvest himself or even to appropriate the already-harvested coffee berries and keep the proceeds. It is obvious that if ex-husbands appropriate coffee income, women farmers are unlikely to tend the coffee plants carefully.

Alternative Cash Crops and Investments

Women, land-hungry youth, or other farmers who reduce their coffee output do not return to a pure subsistence economy but rather invest their labour and capital in alternative cash crop or trading activities. One of the most common alternative ways to earn cash is growing more beer bananas and related brewing and distilling, discussed in the previous chapter. When the productive capacity of older coffee trees declines, some smallholders choose to replace them with beer bananas. (It is illegal to cut down coffee trees in Tanzania without official consent but as is the case with most regulation of smallholders this law is unenforceable on a mass scale.)

Beer bananas are highly resistant to banana weevils and, unlike coffee farming, good yields are possible without an input of insecticide. Pruning, by cutting off the dry outside fibre of the banana tree's trunk, is simpler and more widely practised than coffee-tree pruning, which requires that the farmer cut branches and even tie or graft them. Banana cuttings take a year and a half to mature, whereas coffee seedlings take five years. The cuttings are planted at different times and then yield bananas year-round, while coffee is harvested once per year and the whole crop must be picked in two to three weeks of tedious, labour-intensive effort. Coffee income is paid only once per year, twice if the farmer grows *arabica* as well as the traditional *robusta* crop. The income from beer and liquor, on the other hand, is continuous; the consistent cash flow is used to meet some of the regular cash expenses reported by interviewees: purchased foods such as maize flour, sugar, and fish; school fees and uniforms; and the development tax imposed by the regional government to make up for cuts in central government spending (part of the IMF structural adjustment package, for further details see Boesen et al., 1986).

Mbilinyi (1976) and Tibaijuka (1984) report higher returns for capital invested in bananas instead of coffee. Records from my sample demonstrate that bananas are three times more profitable than coffee; the same ratio as reported by Tibaijuka (1984, p.98).

The limiting case is that coffee can be sold in any quantities, whereas the market for bananas and related products is small and localized. This is why some of the larger farmer are turning to estate production of coffee, where the market and transport infrastructure are assured by the cooperative union or its private competitors. In the past the cooperative guaranteed to buy all coffee at a set price and bear the costs of transport, even if they lost money by this arrangement, however such subsidies have been abandoned. Coffee is usually only chosen as the primary activity on large farms far from population centres, because despite higher transport costs there are fewer alternate ways of investing capital and generating revenues. Only one informant (a relatively rich Muslim trader/farmer) set up a one hundred hectare estate of pure stand *arabica* coffee as a way of using surplus capital. This example demonstrates that even though the mark-up on bananas is greater a farmer with surplus capital may prefer coffee to the small, localized and competitive food crop markets.

The Village Class Structure

The richest 25 percent of farmers in the sample were the ones most consistently increasing their coffee production (despite some constraints) because of their greater access to land, labour, and modern inputs. These farmers, harvested 43 percent of the coffee. Most, purchased land (with capital accumulated in non-farm employment) in addition to their inherited plots; they hired labourers to plant coffee trees, weed and mulch the plots, harvest the coffee cherry, and transport it to the local KCU office.

Richer farmers produce more coffee because they own cows and have access to manure. Under controlled conditions, a one-hectare stand of *arabica* coffee produced 200-500 kilos of unhusked coffee on land prepared without manure and 1,500-2,000 kilos on a plot prepared with adequate cow manure (personal communication from the Director of the Igabiro Farm Training Centre, Kagera region). The upper limit of 2,000 kilos is consistent with the optimal production level on a coffee estate; (Kabwato, 1976, p.96). An interplanted banana coffee plot can produce a maximum of roughly 700 kilos per hectare: this is the level reported by Johnny Bishubi, an interviewee in the ridgetop village and winner of the

Muleba District Farmer of the Year Award in 1985. (Mr. Bishubi's information was likely to be more reliable than the majority of farmers, because he had his ten hectare plot surveyed and kept written records of coffee output.)

Richer farmers wanting to expand coffee production are constrained by land scarcity and laws determining land tenure. (Of the sixty-three richest farmers in the sample, only five derived the largest single share of their income from coffee.) Even if they can purchase holdings, these are rarely contiguous, thus restricting the consolidation of farms and economies of scale.

In 1994 there was an inflow of one-half million Rwandans to United Nations-sponsored special settlements for refugees. The Rwandan crisis in 1994 created a boom market for food and transport especially in the areas closest to the camps. Despite record coffee prices in 1994 and 1995, farmers and traders near the camps were making large profits selling beans to the World Food Programme and plantains to individual refugees and so coffee has been neglected even more than usual. The expulsion of the Rwandan refugees in 1996, may again make coffee an attractive option in the Kagera region, but farms in close proximity to the refugee camps and those suffering from the El Nino floods of 1997/98 must first try to rectify damage to their farms and regional ecology.

Within the sample, the poorer three-quarters (one hundred and eighty-seven farms) produce 57 percent of the coffee volume, but they lack the land, technical skills, and traditional and modern inputs necessary to improve coffee output. Even minimal upkeep of the trees is unusual for these farmers; fewer than five percent reported visits from the agricultural extension officer. About one-third prune their trees, but even a cursory inspection of the plantation shows that most do not prune regularly or effectively. On the poorer farms, many trees are badly overgrown and well past their productive peak, and few poorer farmers (only 26 percent) are planting new coffee seedlings. Rald and Rald (1969, p.45) claimed twenty-eight percent of farmers prune their coffee trees, putting in an average of three days labour per year.

The very poorest households are the least productive and are subject to the worst exploitation. The poorest farmers sometimes fall victim to an usurious arrangement called *obutwara*, described by

Tibaijuka (1979, p.64). The farmer sells —for a cash advance— the right to harvest all the coffee on their plot. The purchaser (albeit unsure of exact prices and the final harvest) offers only twenty to forty percent of the money they may reasonably expect to receive. Four sampled heads of household, all elderly males, utilized this system.

Despite the fact that the Government and the KCU are promoting arabica coffee, the Kagera farmers prefer to stick to their traditional variety, robusta coffee. In Tanzania as a whole, robusta comprises 25% of total output, whereas in Kagera it averages 75%. Farmers prefer robusta because the trees are hardier and require little labour — about one hundred hours per acre or two-thirds the labour requirement for arabica (Mbilinyi, 1976, p.35). Arabica is less disease resistant and requires spraying and more complicated pruning. Its berries fall off when ripe and therefore must be picked in a shorter time period. Even with Government inputs and a fifty percent higher producer price, the farmers of Kagera region are not responding to the call to plant new arabica seedlings. To do so they would lose immediate productive capacity and increase their overall labour requirement. In fact, most small farmers do not even bother to increase production the simple way by pruning and harvesting their out-of-reach berries.

Labour, is the only under-exploited resource available to poor and middle-status coffee farmers. Numerous studies indicate that the village women do a great deal more work than the men, who are not intensively involved in food production (Rald and Rald, 1975, pp. 6, 11, 25, 29; Kamuzora, 1984, pp. 105-124; Swantz, 1985, p.56). The average labour inputs are that male heads of household (in the Bukoba Rural District) worked 2094 hours per year while their wives put in 2801 hours. Women do domestic labour every day as well as farm labour.

There is undoubtedly a reserve of male labour power that could be tapped. Since it is the men who "manage" such crucial aspects of coffee growing as nurseries, planting, pruning, and harvesting, an increased input of actual work on coffee farms could improve the productivity of smallholdings.

The Marketing/Consumption System

The low prices that government marketing boards pay Tanzanian farmers for their produce have drawn critical fire from development analysts, see (Coulson, 1982; Eicher, 1986; Ellis 1982; Lele, 1984). Farmers in the "ridgetop village" expressed overt dissatisfaction in 1987 when the price for robusta was set at 28 Tanzanian shillings per kilo of dried coffee cherry, but were more positive in 1988, when the producer price was increased to 38 shillings and in 1989 when it rose to 51 shillings.

The KCU has succeeded in one important area: its buying centres in the region have managed to issue the farmers their first payments for coffee with minimum delay, unlike the case in past years. The biggest problem with this system is that the KCU financed these payments by borrowing from the National Bank; in 1990 the annual interest rate was 30 percent and has tended to remain at about the same level through the 1990s. Therefore debt service charges constitute the largest single expenditure of the cooperative union, (personal communication Mr. Ishengoma, general manager KCU). It remains to be seen if the private buyers can work out more efficient lower interest credit schemes.

The final payments are made only after the entire national coffee supply has been sold. The purchasing agent, prior to 1995 the KCU, must repay its bank loan and the interest it has incurred in order to pay the first payment; KCU in turn, pays the farmers the final payment, now reduced because of its debt service costs. In 1988, the TCMB paid the KCU 102 shillings per kilo and the farmer received approximately 10 shillings per kilo for the second payment. The smallholder may wait months for the second payment. One Kagera farmer, literate in English, wrote a letter of complaint (published in the Daily News, May 26, 1988) because he never received a second coffee payment: obviously he represents uncounted others who do not correspond with Dar es Salaam newspapers.

This climate of devaluation, low real producer prices, and delayed payments is probably the single most important reason for stagnating coffee production. Coffee payments are best portrayed in reference to what goods and services can be purchased. Villagers use their scarce cash resources to purchase three broad categories of goods: 1) clothing and household goods; 2) farm tools; and 3) food. I was able to reconstruct the order in which households purchase a range of consumer goods. Certain articles such as motorized transport are desirable but beyond the budgets of most small farmers. Other items such as radios are luxuries which a household will forego in order to purchase necessities — goods people will sacrifice a great deal to obtain.

In an earlier publication (Smith, 1987) I was able to measure the popularity of consumer goods. To do this I used a Guttman scale based on an inventory of household goods, supplemented with the following questions: (What goods would you buy if your income doubled? What goods did you buy in the past that you are no longer able to purchase? What goods do you need to improve your farm?). The data from these made it possible to select the best incentive goods: roofsheets, cement and burnt brick, cows, and sugar. This choice of goods is consistent with the findings of a larger national survey of consumer incentives, although some goods singled out by Cooksey (1986) such as rice or cooking oil, were not considered essential items.

The Haya are eager to improve their houses, therefore the demand for reasonably-priced construction materials makes them excellent incentive goods. Despite the high cost of corrugated iron roofing, the vast majority of the 250 households (96 percent) had them; clearly good roofing is a necessity in a zone which receives an average of 1,400 mm of rain per year. Similarly, there is a great demand for reasonably-priced bricks and cement for walls and floors. Even though the prices for locally-made fired mud bricks and factory-made cement were very high relative to incomes, forty-two percent of the sample lived in cement or brick houses.

Cows are another excellent incentive good. Thirty-four percent of the sample obtained cow manure, most collecting it from their own cows even though a local *zebu* or *ankole* calf costs about two-thirds of the median annual family income, and a pure breed dairy cow is at least double this price. Village farmers attach a very high status to cows, mainly for bride wealth and manure rather than their dairy or meat capacity, since local breeds of cattle produce small quantities of both. All the farmers interviewed wanted to buy a cow or to increase their herd in order to improve farm productivity. Cattle are valued both because they are traditional status possessions and because their manure is considered essential to intensive banana and coffee cultivation on small plots. An incentive program to provide cattle or credit to buy them could successfully exploit the high cultural value placed on cattle and their contribution to agricultural productivity.

Despite the high local price of about two dollars per kilo, 68 percent of households in 1990 bought sugar at least once per month. This represents a tremendous financial sacrifice to most families, but sugar is the single most attractive highly-priced non-traditional food.

Coffee production could be bolstered if the purchasers provided the farmer with subsidized sugar, construction materials and cows. They could offer these goods in exchange for coffee or offer a discount price on selected commodities to villages which market coffee in consistently large quantities. The KCU is developing such a program to provide building materials and dairy cows to farmers on credit and to sell sugar at below market prices to villages which supply their coffee to the cooperative.

However, the KCU was constrained by official red tape in its efforts to provide incentive goods, and, by government and party rules; the election of a new government in 1995 has sped up the liberalization process. Bureaucratic interference has not been completely removed by the reforms of the 1980s and 1990s; many of the problems reported by Mbilinyi in the early 1970s still exist. The biggest problem is unclear authority structure: there is often overlap or conflict in the duties assigned to or carried out by the primary societies, the cooperative union, the private sector and the branches of government.

Conclusion

In Kagera Region, the small farmer's primary contact with the national and international economy is selling coffee. Between 1986 and 1994 virtually all smallholders sold coffee, but in a manner consistent with a range of subsistence and cash generating activities in the informal economy.

This chapter highlights why both the Colonial and post Independence states have not been able to significantly alter the small farmer's own strategy of how best to manage coffee as a cash crop. When the state oversteps the farmer's self-imposed limits, the results are resistance — direct resistance in the late 1930s and indirect resistance in the late 1980s and 1990s The Tanzanian state seems to have learned something from the experience of the colonial administration in the 1930s. The massive *ujamaa* resettlement drive carried out across Tanzania in the 1970s was not pursued in the overcrowded Kagera ridges. Here, for the vast majority of villagers, the traditional agrarian organization was allowed to operate on a *de facto* basis. Traditional villages were merely "signposted" and called *ujamaa* villages with no real attempt to enforce the conversion of land to communal plots. Since colonial times Tanzanian governments have been unable to regulate peasant production or to generate substantially higher revenues from smallholders.

Neither the state nor the direct producer can afford a prolonged interruption of coffee production: even during the "riots", actual coffee production was not curtailed. The peasantry was prepared to confront the administration with violence when it attempted to exert direct control over production on the permanent plot, but it was not prepared to give up the "riches and profit" associated with coffee.

By the 1930s the Kagera smallholder had, in fact, become embroiled in a system of coffee production and sale. The real issue behind the 1937 coffee riots was never a threat to return to subsistence production or a refusal to grow coffee (Hyden's "exit option"). Rather, the issue was control over the reproduction of the family farm in a mixed commodity and subsistence-producing enterprise. Raikes (1986) has concluded that in Tanzania resources withdrawn by smallholders from cash-crop activities are most likely to be diverted to the informal sector. This can be confirmed by examining the reported sources of household income for all sub-samples:

1.	trade in primary commodities	54%
2.	salaries and remittances	16%
3.	coffee growing	14%
4.	beer banana, beer and liquor production	12%
5.	sale of food bananas and minor food crops	4%

Coffee growing, trade at official venues, and salaries, generates revenue for the state: the other income-producing activities on the list are centred in the village and are exclusively informal. This sector makes up half the total reported income from all sources. It is significant that the smallholder is "exiting" from coffee production to activities such as brewing, distilling, crafts, salaried employment, etc., which are based on neither the cultivation of crops nor animal husbandry. Political and economic liberalization have not altered the facts of life for peasants.

The only viable way that officials can boost coffee production is to provide a realistic incentives program as a motivation to intensify labour input. Liberalization in the 1990s has helped by making more goods available than in the 1980s however this has not increased aggregate regional coffee production to meet the government's stated objective of improving foreign exchange earnings in order to finance national development. Even record high coffee prices have not arguably, improved the quality of life in mainly poor and middle-status male-headed peasant households by providing access to needed consumer goods and services.

However, the major constraints to longer-term agricultural development outlined in this chapter — scarcity of land, labour, and capital, and constrictions of the gender and class systems — would not be significantly affected by an incentives strategy which focuses only on the coffee cash crop. The interests of landless youth and female-headed households, and of women generally as primary food producers, will not be addressed, since these groups are effectively excluded from controlling participation in coffee cultivation. A strategy based on incentives for coffee production is a viable option for a specific target group, lower-income, male-headed smallholder households. Alternative strategies (for example, those aimed at improving access to inputs for more intensive food production and marketing, non-farm income generation projects, and resettlement schemes) are needed for the expanding population of youth and female-headed households.

9 Sustainable Development Reconsidered: The Rich Farmers

The richest strata of Tanzanian smallholders have already accomplished the objectives of international development. These are usually expressed in terms similar to Schumacher, 1978, p.136:

> ... to help the developing countries to provide their people with the material opportunities for using their talents, living a full and happy life and steadily improving their lot.

They grow surplus quantities of food crops and earn relatively large incomes that they use to purchase a variety of foodstuffs. Barring an unforeseen disaster, their households will always have enough food. They carefully preserve their resources and avoid environmental degradation, therefore, they have learned to "live full lives and steadily improve their lot".

This chapter focuses on two subjects: how the richer farmers accumulate their wealth, and how they sustain their income levels. Like most Haya youth, richer farmers leave their villages in order to earn and save money. The ones who are successful at acquiring capital, usually return to their homeland later in life to set up, or expand, a farm household: the distinguishing mark of the typical richer farmers, is that they make profitable and sustainable investments with the capital they acquired in employment or trade. The richer household usually concentrates on one occupational specialty (to earn the majority of their cash income) but also engages in a diversity of other subsistence and cash generating activities.

Who is a Rich Farmer?

Bader (1975); Rald and Rald (1975); Tibaijuka (1979 and 1984) use fairly simple criteria to define rich farmers. As will become evident in this work the multi- faceted nature of the economic activities pursued by the rich farmer makes a more complicated operational definition necessary. Therefore I used five criteria to demarcate rich farmers:

1) They must have a minimum cash income equivalent to $1250 US (at the exchange rate in effect on the interview date).[1] Their median cash income is just below $2000 US.

2) They must own at least one farm plot, although in some instances they were absentee owners or did not derive much cash income from this plot. The replacement cash value of the average rich farmer's two hectare plot is $3000 US, based on 1988 prices.

3) They must own an improved house, a dwelling made of cement or burnt bricks and covered by iron roofsheets. Houses were the single largest cash investment of most households, an investment justified by their symbolic status value, the obvious comfort they offered, and their real value — as opposed to the soaring rate of inflation for construction materials and house prices. The average sale price for one of these improved houses was the equivalent of US $5000.

4) They must possess most of the status symbols, popular luxury consumer durables which signify security of wealth. These include items such as a sofa set and a radio-cassette. At 1990 prices this represents a further $2000 US expenditure.

5) At least two other household heads in their respective subsample must have singled them out as rich.

Sixty-three of the two-hundred and fifty selected households met all these criteria. This richer group owned a disproportionate share of cattle (the traditional status symbol) and they monopolized motorized transport, the ultimate modern status symbol: they owned all the cars, pickups and lorries, as well as, sixty nine percent of the cows. Not all rich farmers own livestock or a motor vehicle but the mean value of these goods for the rich farmer added another $3000 US capitalization. The average rich farmer in this sample had also accumulated durable goods with a value equivalent to about $13,000 US — almost seven years of their present average annual income. (Durable goods refers to houses, land, livestock, consumer goods, vehicles and means of production.)

Their situation contrasts markedly with the poor farmers, who I defined as those who have insufficient land (less than two acres) for the household (three or more persons) and a small cash income (less than $500 TSh equivalent per year). Middle farmers are defined as the residual category. Using these criteria the basic socio-economic characteristics of the groups are presented in the table below.

Table 1: Select Characteristics of Head of Household and Family by Socio-economic Status

	Rich	Middle	Poor
Age of household head (median)	44	43	38
Percentage female-headed	8%	27%	33%
Permanent residents (median)	5	4	4
Non-resident children (mean)	2.4	1.1	0.5
Royal clan membership	25%	27%	23%
Formal education (median years)	8	7	7
Religion			
Catholic	54%	50%	48%
Protestant	24%	28%	39%
Islam.	21%	21%	13%
Previous out-migration (percentage)	87.5	52%	35.5%
Farm acreage (mean permanent plot(s)	4.5	3	1.2
Total Cash Income (median US Dollars)	$2,748	$582	$350
Membership in an official political party	82.5%	57.5%	45.2%

The richer farmers can all be described by adjectives such as clever, well-organized, hard working and innovative, but these characteristics apply equally well to many of their poor and middle status compatriots. What is the crucial sociological component in their successful careers? I believe the career patterns of these richer farmers can be used to test the utility of five sociological approaches to international development: modernization theory, peasant economy, dependency theory, gender differentiation, and the accumulation of capital. The varying predictive power of each of these five approaches, neither proves nor disproves the theory, but the exercise does highlight a primary difficulty in the social sciences: concrete case studies are rarely exact reflections of patterns that sociological theory predicts.

Pathways to Wealth

(1) Modernization Theory: Ascription and Achievement

Modernization theory assumes that individual social and economic advancement will depend mainly on ability and achievement in a modern society or inheritance and ascription in a more traditional one (influential texts include: Eisenstadt, 1974; Levy, 1967; Huntington, 1976 and 1987; McClelland 1964; Parsons, 1951; Rostow, 1971 and 1990.) Talcott Parson's "pattern variables" provide the central principle in modernization theory (see the excellent analysis by Savage 1977 or So 1990) and by Parsonian logic, the richer farmers should be either more modern (and presumably better educated) or those who have inherited their positions in the more traditional pattern.

For this sample, neither exposure to modern ideas in the school system, nor the inheritance of family position proved crucial in determining who becomes a rich farmer. The median education levels of the rich (8 years) versus the poor and middle status group (7 years) is not significantly different and there is no meaningful correlation between years of schooling and total cash income. For certain individuals of course, education led to upper level jobs and access to savings and scarce capital resources such as bank loans. Within the Tanzanian state system, salaries are low even at the upper levels, so the lack of correlation between years of schooling and income is not surprising.

Many parents thought education to be extremely important and some made great sacrifices to educate their children. The vast majority of informants who attended Secondary School came from families in which most of the children had attained the same level of education, therefore education was most strongly influenced by modern family attitudes and the ability to pay school fees and forego family labour. But education did not lead inevitably to wealth and status.

The richer group *were* more knowledgeable and sophisticated farmers: they were more than twice as likely (40%) than the low/middle group (17%) to have taken a course on improved methods of farming or consulted extensively with the Agricultural or Veterinary Extension workers. This was the *result* of the need for advice on running larger farms and in the case of those who had bought improved dairy stock, the course was compulsory before they could obtain their cows; however this in no way *proves* that richer farmers had become more "universalistic and modern" than their "traditional and particularistic" neighbours.

Barth's classic work on peasant entrepreneurs (an adaptation of Merton's social theory) posits that entrepreneurship is the critical variable in development; a conclusion consistent with David McClelland's classic formulation that "achievement motivation" exemplified by domestic entrepreneurs is the key to economic modernization (So, 1990). According to Barth peasant entrepreneurs often ignore the normative structures of the community to further their economic interests, these "free enterprisers" tend to speculate and disregard community social values (Firth, 1967, p.63). Since the activities of the richer farmers do not pose a severe social or economic threat to existing norms, I see no evidence of a "free enterpriser" type emerging. On the contrary, Haya communities today, as a whole, readily accept profit-making innovations.

Religious background does not appear to be a significant factor in predicting an achievement orientation and upward social mobility. I completely agree with another researcher (Stevens, 1991) who highlights the difficulty in distinguishing "Protestant", "Catholic" or "Muslim" behaviour within the context of the Haya village. Although the Protestant churches were the only ones involved in promoting birth control, there is no consistent evidence that Protestant villagers were more likely to give up traditional practices or exhibit greater entrepreneurship, innovation, or a superior work ethic.

One Protestant individual who did work his way up from a very low social position to become a rich farmer, owes his success more to "clientage" (which Bryceson, 1990, identifies as the prevailing pattern in Tanzania) than modern beliefs. The Evangelical Lutheran Church gave him his first job and then trained him to teach bible classes, it also supplied him with a motor vehicle from the mission headquarters in Bukoba. It is clear that missionary assistance can be an important factor in individual social mobility, but there was no trend for Catholics, Protestants or Muslims to acquire greater wealth. The proportions of the three largest religious groups (Roman Catholics, Lutherans, and Muslims) is the same in the richer and poorer subsamples.

The "traditional" system is supposed to be governed by ascription and inheritance: but in Hayaland, social status cannot be predicted by ascribed characteristics. Inheritance, was no more important in acquiring wealth and status than education. These concepts explain only a few individual cases. A good example of ascribed privilege and position was the trading monopoly exercised by Asians and Muslims of Arab descent during the colonial period: these privileges did not survive after Independence when restrictions were lifted and Africans took over most trading. The vast majority of sampled rich farmers, engaged in trade as a primary occupation, are Catholic.

Inherited wealth was an important factor in one individual case: that of the richest man in the sample. As the son of a Muslim trader, his accumulation of wealth began with an enormous advantage, the inheritance of a well stocked shop and a lorry.

The clan system, no longer denotes access to privilege as it did in the past. For about three hundred years prior to Tanzania's Independence, the Bahinda clan controlled the state and productive property (arable land and cattle). The Germans and the British recognized the hegemony of the clan system and allowed the Bahinda to remain as the acting rulers in the region. Despite the longevity of clan rule, the vast majority of Bahinda have not retained their privileges. There is no difference in the proportion of Bahinda (about 1/4) in the richer and poorer groups. Of course, there are individuals who have retained inherited privilege. The heir to the former chief of Kiziba District, owned a large well-managed coffee plantation, a car and an improved house. In this case, land inheritance provided security for a bank loan which allowed him to expand and improve his farm. But for most Bahinda, their large families and consumption ethic prevented them from preserving their privileges after Independence and the abolition of their rule. The cases of the Bahinda and the Muslim traders demonstrate that "the economy of affection" does not necessarily transmit its privileges from one generation to the next.

(2) Peasant Economy and the Family Farm Cycle

The term "peasant society" often denotes a system mostly isolated from the world economy and its socio-economic connections. Raymond Firth provides the classic definition of peasant economy:

> ... a system of small-scale producers with a simple technology and equipment ... a small-scale productive organization built upon a use of or a close relation to primary resources, has its own concomitant systems of capital accumulation and indebtedness of marketing and distribution ... we can speak not only of peasant agriculturalists, but also of peasant fisherman, peasant craftsmen, and peasant marketers ... such people are part-time cultivators as well. (Frankenberg, 1967, pp.52-53)

In purely descriptive terms this model does indeed apply to the Haya. But peasant economy theorization fails to provide a reasonable structural base to understand the observed realities of the Kagera village, because of its central assumption — that land is a free good. Goody (1971) portrays Africa as a continent where land is usually plentiful, having little if any economic value. Arable land on Kagera region's densely populated ridges has been scarce since colonial times and with Africa's rapidly growing population, land scarcity is becoming a more generalized phenomena.

Labour is supposed to be the limiting factor of production, wealth increases as the size of the family increases, because there is more labour to work more land. Wealth is a function of family size, following Chayanov's basic idea of the labour/consumer balance. The "peasant economy" approach presupposes that the richest households must have the largest number of working wives and children. This was not the case for my interviewees, although there is a difference between the average household size of 7.2 for the richer and 4.9 for the poorer groups. However these numbers include non-residents and the richer group has an average of 1.5 more outmigrants. This could, in theory, signal that wealth depended on remittances from migrant children. However, the amounts of money actually remitted were too small to make any real difference in farm outputs or standard of living indicators. The richer group received a mean remittance income of US$57 while the poorer group received US$34, a small difference in relation to overall incomes.

It would be impossible to predict which sampled households are members of the rich farmer group knowing only the total number of persons in the household or the age of the head of household. (The median age of the head of household and the median family size are only slightly greater for the rich group, see table 1).

The rich farmer group controls about double the per capita arable land, but their access to more land was not the result of large families working free land or the result of inheritance. The same proportion of heads of family had inherited land from both the rich and poor groups: but 77% of the rich group and only 24% of the poorer group had purchased at least one additional plot — cash is the prerequisite to obtaining more land.

In this study, the classic family farm cycle is most important in the Ujamaa village, where in the late 1960s and early 1970s, free land grants were made to adult members of the village corporation who worked on the communal plots. Households with several adult members could acquire more than one plot, provided they developed their land grant. The combination of free land incentives and a poorly developed cash economy enabled those with larger families and extended family assistance to develop more agricultural land (Bader, 1975). By the late 1970s, land grants were no longer available and developed land was being sold at regional market values. After selling their surplus land, those with the wherewithal could reinvest profits in more lucrative but more highly capitalized ventures, such as pickups, dairy cows, shops, fishing nets and outboard motors.

Mathias' life story illustrates this pattern of deferred gratification and accumulation of free land. He had worked as a "houseboy" and watchman in Bukoba town in the 1960s, saved as much as possible and then migrated to the Ujamaa village. With seven resident workers, two wives and five children, the family was able to develop four plots of free land by using family labour supplemented by part time occasional employees. By the late 1970s he sold two plots and reinvested the proceeds in a general store, which is now his primary source of income.

This life history is also typical of a family farm cycle in which the head of the household must work off the farm in the cash economy until he or she has accumulated enough capital to return to the farm-based enterprise. This pattern differs from that reported by Reining in the 1950s because at that time the occupational specialties did not require out migration to towns in order to accumulate cash for entry level farming. Boesen et al. (1977, pp.25-30) reported that even by the early 1970s the family farm cycle based on land inheritance was rapidly breaking down.

Most of the internal migration literature focuses on the problem of rural-urban migration and surplus urban population. Todaro (1976, p.9) for example, points out that Tanzania's average annual population growth rate between 1948 and 1967 is 2.5%, but 6.8% in the urban areas where 84% are migrants. The evidence from Kagera region indicates that migration to urban areas is not permanent: urban migrants aspire to become rich farmers, most return to the rural areas after they retire or accumulate a substantial amount of cash.

Whether or not this scenario becomes more common elsewhere in tropical Africa, in Northwest Tanzania, out-migration is the typical pattern for a successful farmer. Only two of the richer farmers — those who derive the major share of their income from coffee farming — had not left their home villages for any extended time. They are older by a median of eleven years than the rich farmer average and they enjoyed special advantages (inheritance of large plots). Their career cycles are characteristic of the 1950s and 1960s, which is precisely when they inherited their land.

The current cycle of out-migration and return is exemplified in the life-histories of the majority of rich farmers, who had to migrate to accumulate farm capital. The case of the ex-houseboy is only slightly different than most because of his access to free land.

The practice of polygamy *was* more prevalent among the richer farmers. Of the 250 households, twenty eight reported polygamous marriages, and eighteen of these were in the richer strata. Access to additional labour is undoubtedly a benefit to the polygamists, but the interviewees in this survey were already rich before they acquired a second and in eight cases additional wives.

(3) Dependency Theory and Africanization

Underdevelopment and dependency theory assumes that most economic behaviour in the Third World is conditioned by the centre/periphery relationship. Official development assistance breeds dependency, corruption, economies based on mono-production export crops, unequal exchange, low real prices, and externally imposed conditions of structural adjustment. These conditions are labelled "underdevelopment". In Tanzania, where two-thirds of the Gross Domestic Product of the 1980s came from foreign agencies, the national economy was certainly highly dependent.

In the literature about Tanzania, Michaela Von Freyhold is the most sophisticated proponent of a dependency type of analysis. She portrays four *primary* classes operating within the Tanzanian national context:

- the ruling class are the agents who dispense development assistance,
- the governing class is the state bureaucracy or Nizers (from

Africanizers),

- the rich peasants or "kulaks",
- the masses of the peasantry. (Von Freyhold, 1977 and 1979)

The nizer class depends on external support (Von Freyhold, 1979, p.115) therefore nizers really represent the interests of external powers, whose mission is mainly to supply the world market with tropical raw materials at favourable prices. The nizers ability to carry out this function is the guarantee that foreign aid will continue and once the infrastructure is in place foreign investments are supposed to follow. In return the nizer group consolidates its power, privilege, and conspicuous consumption. (Bolton, 1985, p.166)

In Von Freyhold's analysis, the rich peasant is above all an agricultural producer, whose basic interests are in opposition to the nizer bureaucracy who support their foreign masters (Von Freyhold, 1979, p.63). But at the same time, the rich peasant is an exploiter of cheap labour, a profiteering merchant trader, and above all a land grabber.

One problem with Von Freyhold's demarcation of "classes" is that often the bureaucrat and the rich farmer are the same person at different points in their life-cycle, as the nizer becomes a kulak after official retirement. This trend is very clear in the Kagera Region.

In this study, I defined "nizer" as an official of any of the following bureaucracies: the government, political parties, parastatals, semi-official institutions (such as cooperative unions) and parallel bureaucracies (such as private hospitals and schools) — in short, the technocrats who took over after Independence and Africanization. I observed four predominant nizer-type occupations, these are: 1) *officials* (people in leadership positions in the civil service, the ruling party and the regional direction of the cooperative union); 2) *managers* (upper level and middle level, accountants and managers in the extensive parastatal sector); 3) *scientific* (technicians, laboratory, medical, veterinary, agricultural extension, and engineering personnel); 4) *teachers* in primary schools, secondary schools and colleges.

These titles exaggerate the formal qualifications, and status of the people in question. In the Tanzanian context, teachers, medical assistants and so forth are often recruited on the basis of their ability to learn practical skills on the job or in colleges which they may enter after primary school.

Nizers, in the period just before multi-party elections, were all members of a political party. (Despite party membership few are politically active). In contrast, 82 percent of the rich farmer group as a whole and 62 percent of the poorer subsample reported official party membership. Since a relatively large proportion of nizers work for the state, this calls into question the notion that civil society tends to be autonomous from African governments.

Nizers, as defined above, were the largest single occupational cluster to *become* rich farmers: forty six percent of the "kulaks" were once "nizers". In the career cycle of a typical rich farmer work at an official post was a transitional stage which assisted the child of a middle peasant to become a high-status farmer. The key to the transition is the accumulation of a critical mass of cash. The rich farmers who enter this strata from the nizer group have one big advantage over the others, they have a higher success rate in obtaining bank loans. Fifty-nine percent of all bank loans made to rich farmers went to this sub-group, even though they represented only forty-six percent of rich farmers. Only six percent of middle status farmers and no poor farmers obtained bank loans.[2]

Nizers use the official economy to establish themselves as rich farmers, but once established their interests are usually opposed to the bureaucracy's quest for foreign exchange from export crops. The primary corroboration of Von Freyhold's class analysis is the rejection by rich farmers of the government's "grow more coffee" campaign discussed in the previous chapter. Cash incomes derived from coffee, represent only 14 percent of the total within the sampled 250 households. The rich farmers, on average, produced twice as much coffee as the others but only a slightly greater *proportion* of their income came from coffee. This is not because the villages I selected lie outside of Tanzanian participation in the world economy, in fact the opposite is true. Within the Tanzanian context, the Bukoba and Muleba districts are two of the most important coffee producers in the country; coffee is the nation's number one export crop and Kagera's production represents about 20 percent of national exports by value (Tibaijuka, 1984, p.50; ERB, 1988).

It is significant that the largest coffee growers in a key area derive less than one-fifth of their total cash income from the nation's number one source of foreign exchange. This indicates that the structural opposition between nizers and farmers may be less important than it seems to Von Freyhold, since it is grounded only in the official economy.

The relative decline of coffee corroborates my basic assumption that the real path of personal accumulation is "articulated accumulation," the production and sale of foodstuffs and other locally produced commodities (O'Brien, 1986, p.199). "Nizers" use official positions to accumulate their base capital and then (later in their life-cycle) draw most of their income as rich farmers from the parallel economy.

As an illustration of this pattern, consider the career history of one of my informants. FJR was born into a non-royal middle status family in the late 1920s, the coffee boom period. He completed his primary and secondary schooling during World War two. The Balimi fund, a program of the regional Cooperative Union, provided a scholarship up to Form Four. In the late 1950s he won another scholarship to study accounting in Dar es Salaam, and on graduation was employed by the Colonial Government. After Independence, in December 1961, FJR was employed by the Tanzanian Government in a series of high-level posts including, District Officer and Town Director in Singida, planner in the President's Office, Regional Commissioner in Iringa, and District Commissioner in Mwanza. His government service lasted until 1979 including a two year paid leave of absence to study abroad. At retirement he obtained a loan worth US$ 10,000 from the National Bank of Commerce to build up a progressive type of farm enterprise. He became one of the richest men in the District and owned five farm plots, three cement houses and two lorries: about one-half of his yearly cash income came from transport. Like most other rich farmers, he owned a series of enterprises, which he ran with the assistance of twenty hired full-time workers and occasional part-time workers, such as coffee pickers. His income sources included: banana/coffee plantations, which produced about two tonnes of coffee and surplus plantain and beer bananas; a dairy with ten purebred cows that produced about one-half million litres of milk per year; a market garden producing tomatoes and other vegetables for the Bukoba market;

a poultry farm producing 160 eggs per day and 10 broilers per week; and a woodlot from which FJR harvested firewood and lumber — cut by workers using the handsaw and pit method. His life history seems to exemplify my assertion that the nizer/kulak roles are complementary in the career cycle, even though they may be structurally opposed to each other as Von Freyhold asserts.

The nizer pattern is being modified because of structural adjustment, devaluation and high rates of inflation, all of which reduce official salaries to insufficient levels: as Sandbrook (1993) points out this is happening throughout Africa. As essential services break down. Nizers hold on to their jobs more because of perks than salaries, subsidized housing, travel allowances, trips abroad, official cars, etc., are often worth much more than salaries. During my stay in Tanzania few mid or upper level technocrats spent much time in their offices. Postal officials, customs agents, motor vehicle staff, university administrators, teachers, lecturers, police officers, civil servants, etc. all ran independent side businesses such as dairies, chicken and egg production, transport using pickups or fruit plantations. The "nizers" of the 1980s and 1990s were "kulaks" even while they were occupying official positions. Since Tanzania's managerial strata were drawing salaries but not putting in many hours; the spin-off was of course a very inefficient infrastructure.

Clientage throughout both the state and the economy may explain the poor performance of African bureaucracies and the need to improve instrumental rationality. However it is premature to look to the informal economy, civil society and the middle classes as agents of transformation, transparency and accountability. Our examples show clearly that top level agents in the informal economy, the supposed leaders of civil society are connected through the career cycle to the governing classes. One must surely question assumptions that middle class accumulators, are independent of state run bureaucracies and patronage networks. For example, Bratton (1994, p.58) assumes an unproven generalization true mainly in the urban enclaves:

> Middle class elements are prominent in founding and leading civic organizations and in articulating 'universal' values as a

means of countering particularistic loyalties and building broad, multiclass political coalitions

(4) Accumulation of Capital

By diversifying rather than by growing more coffee, the rich farmer escapes the trap Bernstein (1979) refers to as the simple reproduction squeeze: the result of concentrating too many resources into cash crops subject to low official prices. To escape the "squeeze" the farmer must accumulate cash, land and other capital. He, or infrequently she, then enters what Marx described as the "circuit of capital": money (M) is invested in commodities (C) for trade or production and eventual sale for more money (M') or in symbols (M-C-M').

Besides the nizer route, there were a range of other occupations which also could lead to the initial sum of cash for investment and the subsequent flow of money income. These occupational categories are:

- *fundi*, a Swahili word (the nearest English equivalent is skilled craftsmen);
- salaried work in unskilled and semiskilled occupations;
- a residual category including fishermen, traders, coffee farmers and prostitutes.

Twenty percent of the richer farmers accumulated their start-up capital by work as *fundi.* With existing structural adjustment programs, low official salaries and increasing competition by those with the requisite training to enter the bureaucracy, the *fundi* route could become even more important in future. Skilled craftsmen are extremely useful because they produce and repair high utility commodities. This career pattern can best be illustrated with an example. The most successful *fundi* in the sample spent his early years assisting his father in constructing rectangular dried-mud houses, he then left his home village to pursue secondary studies in Bukoba town, where he lived with his uncle, a carpenter. After two years he left school and worked as a full-time assistant to his uncle. Five years later, he was able to buy tools and open his own small workshop. He learned furniture making and some basic upholstering, and then began a very successful artisanal enterprise producing wood frame sofas, armchairs and replacement car seats. In 1989 he earned (in Tanzanian shillings) about three hundred US dollars per month, more than three times the highest official salary recorded in this study.

Drivers are an important *fundi* specialty (they must also know enough mechanics to be able to deal with common breakdowns of their used vehicles). A driver usually learns his trade from older kin or friends, and then will be hired on the recommendation of other drivers. Drivers need stamina for gruelling long distance transport, but if they avoid motoring accidents, they can become rich. Drivers were the richest occupational group in the sample. The five individuals who derived the largest share of their income by transporting goods and people, had the highest median cash income — four thousand dollars per year. Due to the scarcity of transport and poor roads, lorry owners, as well as hired drivers, are also in a favoured position to exploit the classical circuit of merchants capital. Drivers leave Bukoba or Mwanza with relatively cheap manufactured goods which they can sell at a profit in the villages, where they purchase a stock of cheap agricultural goods to re-sell at a profit in the towns. Drivers can also load up their trucks with paying passengers. One of my informants, a driver for a large parastatal trading company, admitted he was making two thousand dollars per month from his unofficial trading. After the genocide in Rwanda in 1994, one-half million refugees fled to UNHCR-administered camps in Kagera region. Food had to be trucked in and personnel moved about; there was much work for drivers. Drivers for the relief agencies were better paid than those working for official Tanzanian agencies although they had fewer opportunities for private profit. Nevertheless, I was able to reach a remote refugee camp in January 1995 by paying an unofficial taxi fee to a Red Cross driver whose vehicle was "being serviced".

Unskilled or semi-skilled salaried work provided the start up capital for nine rich farmers. I have already described the career of the former houseboy who moved to the *ujamaa* village to take up free land

and eventually to become the owner of a general store. In his and a few other cases, only a small amount of start up capital was required as the initial base. Poor peasants who harvest tea on the estates, or those who pick other farmer's coffee, or who migrate to the towns to search for menial jobs, have not developed a class consciousness, because they aspire to own land (or more land) and many of them believe they can emulate the few rich farmers in each village who started small. Von Freyhold (1979, p.65) points out an analogous inhibition to class consciousness among wage workers in the Handeni district.

However values usually reflect material conditions and these are changing. In 1990 and again in 1994, I observed a phenomena I had never seen before in Tanzania, a visible underclass of hundreds of homeless street children and adults, sleeping on the pavements of Mwanza, Tanzania's most economically dynamic metropolis.

The challenge facing an out-migrant (a young person immersed in the cash economy) is to save up a reasonable amount of start-up capital to invest in trading or a service enterprise. One-third of the rich farmer group derived their single largest source of income from trade, and many of them did indeed start small. In using the term trader I include: commodity traders, owners of a *duka* or general store, and sellers of used clothing. If owners of illegal bars, food sellers, owners of *hoteli* (small restaurants) and caterers are also considered traders, then this category would include one-half of the rich farmer group.

Trading forms a chain of links between producer and consumer. The majority of farm families in the Kagera region live outside of the commercial centres, they have low incomes, little capital and storage capacity and they must buy in small lots. Bauer neatly summarizes the prevailing conditions,

> ... consumers wish to buy or have to buy in very small individual lots. It is expensive to supply in this form because of the cost of transport and storage, and the necessity to break bulk many times. Thus the tasks of distribution in Nigeria and similar economies absorb a large volume of resources. (Bauer, 1974, pp.16-17)

Distribution of commodities takes place in many stages: it is labour intensive and creates employment whether or not domestic or imported goods are being traded.

Traders (in the sample) purchased food, firewood or agricultural items in lots they could sell at a higher unit price. Usually, this type of operation involves transporting goods from one place to another. Wycliffe travelled twice per month by ferry from Bukoba to Mwanza, where he purchased fifty-kilo gunny sacks of maize for resale in Bukoba town. His profits from this enterprise were two hundred dollars per month or double the maximum income anyone in the sample earned from coffee sales.

Roadside or market stands, used clothing sales, catering, *konyagi* (banana liquor) bars all represent areas of endeavour where it is possible to start with a very small amount of capital. Such enterprises are entirely self financing: not a single respondent whose most profitable activity was one of the above obtained start-up capital from a bank loan.

Mitumba, or second hand and remaindered clothes, are sold in even the remotest villages. These come in sealed bales which are opened and sold piece by piece. The richer dealers usually buy bales and open market stalls, but one of my informants explained how she selected pieces from the central market as soon as the stalls opened for business, and then sold them in outlying villages. Mission connections help in this field, since the bales sold by church groups are offered at lower than market prices to those active in the church.

Ownership of a *duka* or general store usually brings in a high and reliable income but few can afford to purchase the initial inventory: only seven individuals or eleven percent of the rich farmer group derived their largest income share from a general store. Of the seven, three had inherited their position, starting off their working careers as shopkeepers. The risk, difficulty and expense of setting up a well-stocked, well-managed general store make it a minority occupation which like transport can be very lucrative but very difficult to gain access to.

(5) Gender

Gender plays a major role in acquiring riches. Only five of the sixty households headed by women had high enough incomes to be considered rich and all of these were at or close to the minimum cutoff point. Since so few women became rich farmers, I will save the discussion of these five cases for the next chapter on female-headed households. The life histories of these unusual women give credence to development theorists such as Guyer and Swantz who document how the restrictions of patriarchal inheritance and other patriarchal cultural patterns are such an enormous obstacle to development and a waste of human potential.

Rich Farmers and Development

In the 1950s the British Colonial Administration inaugurated a development policy explicitly designed to assist rich farmers, Coulson (1982) refers to it as the focal point approach. The basic assumptions were that the more enterprising and successful smallholders would both set an example and act as leaders in establishing and extending income-generating projects. The British promoted coffee and other cash crops as a means to foster economic growth, infrastructural development, and in turn bring more wealth into colonial coffers. More smallholder wealth generated more revenue for the Government through export taxes and import duties. Nevertheless, development in any meaningful sense, was limited by British unwillingness to build up an extensive infrastructure in Northwest Tanganyika.

Does a "focal point" type of development strategy necessarily imply "betting on the strong" and therefore undermining the weak, as Bernstein (1977 and 1981) asserts? The richer farmers in this study all manage sustainable enterprises, which operate at a scale appropriate to local conditions. They are largely self-financing usually raising their own start-up capital Rich farmers boost local enterprise because they are the largest consumers of local materials: they provide credit, loans and other financial services in the villages and supply a range of goods and services which villagers demand. The marketing process is by necessity labour-intensive, since in the vast majority of transactions the producers sell and the consumers buy small lots of goods at frequent intervals. In this cash-poor economy with its poorly developed transport and communications infrastructure, its dispersed consumers and producers, there is a need for numerous intermediaries. Petty traders may derive a relatively high commission from middleman activities, however they fulfill a function which cannot be replaced by alternative organizations based on economies of scale. It is not surprising that Tanzania's attempt to nationalize small shops in 1981-1982, and have their activities taken over by the official Regional Trading Corporations, ended in abysmal failure.

Unlike many larger projects dependent on development assistance, rich farmers pay their recurrent expenses from their own profits and set a positive example for others to emulate. Could national and international development workers fond of claiming their work aims to "alleviate poverty" extend the rich farmer's initiatives and use these as a foundation for further development? The rich farmers have pioneered more productive methods of food production, alternative cash crops (including food crops and swing crops), viable transport and marketing networks, and successful cottage industries. Could such methods be adapted to serve the needs of the poor — landless or nearly landless youth, female-headed households, AIDS orphans, and others who have so far had little chance of accumulating capital and providing themselves with sustainable incomes?

Will it be possible for the poorest rural dwellers to obtain appropriate technical assistance, modelled on the innovations of their richer peers? They need training, rural credit schemes, demonstration projects by actual producers, and incentives to trade, such as inexpensive covered market stalls. Successful projects operating at an appropriate level, would generate revenues which could be channelled into infrastructural development and better health and sanitation initiatives. If the rich farmers are to provide a positive model for development then care must be taken not to stifle initiative and enterprise but at the same time to set limits. Kagera Region, Tanzania, need not turn into a

10 Women Heads of Household: The Need for Empowerment

Poverty and patriarchal[1] custom are the principal obstacles women face in trying to set up independent households.[2] Haya women are the most productive members of their society but paradoxically are the poorest. This study of four villages in Hayaland, confirmed a common tendency in rural Africa: women carried out approximately two-thirds of all work but only earned one third of the aggregate cash income. Most of their meager incomes was used to care for children leaving little savings or capital accumulation. Women, out of necessity, have developed a work ethic and skills at managing their meager resources. This chapter explores the main reasons why women are disadvantaged. These women's resourcefulness suggests the need to increase their power as consumers and decision-makers in order to promote regional development.

In 1995, Tanzania held multi-party elections and continued to liberalize its economy. Despite multi-partyism and their geographical and cultural distinctiveness, the Haya have not developed an ethnically based civil society (associational life interacting with and influencing the state). There are separate women's associations of which the informal *kyama* is the most important.

The fieldwork for this research was carried out between 1986 and 1996. I used a sample of two hundred and fifty households drawn from four villages. I was resident in a village of 1200 people for a period of ten months and then returned six times to the region for varying periods totalling a year. To date, there is little empirical data on the type of independent women farmers who make up three out of ten of surveyed heads of households in this study. To help fill in this gap I interviewed a sample of seventy six female-headed households and one hundred and thirteen wives, inquiring about land holdings, crops and yields, tools, inputs and methods of cultivation, division of labour, sources of income, household expenses, health and nutrition.

Forced Migration and Gender Constraints

Kagera's population tripled in the past forty years. Haya sons and daughters were often forced to set up new farms on lands that villagers considered *marginal* — soils needed to be upgraded, bush had to be cleared, tree crops planted and frequently the location was far from markets. Along with expansion onto marginal lands came another major adaptation to population pressure, out migration. The two processes were complementary since for the Haya the most common method to accumulate savings, both for the purchase and development of arable land, was to work off-the-farm for several years. Marginal land *can be* made into profitable and productive farms but this requires an injection of capital.

A larger proportion of women migrate to accumulate capital. The essential problem for Haya female out migrants is that they tend to have less education and fewer marketable skills, however, they *need to* accumulate capital because of gender constraints. Sampled households reported that 60 percent of sons and daughters between the ages of 25 and 45 migrated to urban areas or to full-time non-farm employment. This is the same as the proportion of interviewees who had themselves resided outside their villages. Sampled female and male migrants spent seven to nineteen years outside the region. Seventy percent of women (fifty-three out of seventy-six households) were returned migrants, whereas, only fifty three percent of male-headed households had migrated (ninety-six out of one hundred and seventy-four). Women and the richer male sub-sample exhibited a similar pattern (Smith, 1994): nine out of ten of richest male farmers migrated in order to earn and save. However, most women work outside in order to return and survive and rich farmers return to prosperity.

It is not clear what proportion of the current out migrants will return. The pull factors (jobs, relationships, careers, business opportunities) and the push factors (overpopulation, land scarcity, environmental degradation) make this an open question — one that will need to be researched. For women the alternative to out migration may be marriage. Women who marry and stay in the region gain social status but they work very hard and seldom are able to claim land or other wealth should their husbands die or divorce them.

Out Migration and In Migration

Out migration from heavily-settled areas of Kagera, such as Bukoba and Muleba districts, will increase unless alternatives such as population control, rural industries, or intensive farming programs are provided. Since over 60 percent of residents are under 25 years of age — and the average sampled household consists of 5.2 persons — the population could double every fifteen to twenty years, faster than projection of twenty-five years for Africa as a whole. Given the existing farming system, the region cannot produce enough food to sustain its population and in future out migration may take place under increasingly desperate circumstances. Village poverty augments the number of female-headed households because there is a shortage of men able to pay bride price and start a family farm. Divorce is another social problem that augments the number of female out migrants. Men who divorce their wives sometimes permit ex-wives to remain on the farm grow food and provide subsistence to the children. However, in most cases, divorced women are required to leave the farms they have worked hard to develop; if they are to survive on their own they will need enough cash to buy a small plot.

Unlike the migrants of West and Southern Africa who tend to reside permanently in urban centres, the Haya more often migrate for a period of years but then return. Out-migrants are of two types: 1) those with the best education and training able to find good jobs in the public sector or start successful private businesses — the most successful of these migrate to the commercial centres, Dar es Salaam, Arusha, Mwanza; 2) those who are landless and poor (usually badly educated) who migrate merely to survive, a greater proportion of women fall into this category. Unskilled migrants face a difficult and sometimes desperate situation but it is not always hopeless or impossible; some manage to find jobs or business opportunities and accumulate capital for their return. One-half of female out migrants were government employees, teachers, nurses and similar white collar workers. The remainder were homemakers, service workers or students.

Around ten thousand Tutsi from Rwanda established themselves as self settled refugees in the 1960s, 1970s and 1980s. However after the civil war in Rwanda in 1994, a cataclysmic migration of Hutus took place. About one-half million Rwandese Hutu fled into the Kagera region — this is the equivalent to one-half the local population. Rwandan settlement was confined to refugee camps administered by the UNHCR. Even though refugee camps were all located in the underpopulated border regions they generated a great deal of resentment and opposition at the national level. In 1995 I monitored the major East African newspapers: I did not see a single positive statement, about Rwandan refugees, by any Tanzanian politicians, leaders of national associations or other opinion-makers. In the densely populated Kagera region, there was no room for a half million refugees even in the underpopulated border districts. The consensus reported in the press was that Hutu refugees were an economic burden, a security threat, and the camps were an environmental disaster. By 1997 the government had closed down all the camps and repatriated these refugees back to Rwanda and Burundi. Obviously some refugees, including some guilty of crimes against humanity, must have gone underground. However, these individuals would have migrated to urban centres far from the sampled villages in Kagera region.

Some village residents profited from the Rwandan crisis: the relief effort provided a boom market for local food surpluses. Since male-headed households are larger, they sold more food and reaped more benefits from the Rwandan crisis and relief efforts. Security problems, and the need to employ large numbers of local drivers, construction workers and security personnel tended to reinforce the existing gender bias in employment and income distribution; a much greater proportion of males either worked in the camps and/or were able to take advantage of trading opportunities. The end result of the Rwandan refugee settlements seems to be an intensification of the existing inequalities. The situation reported by Tibaijuka (1984, p.77) has persisted; males predominate in: "wage employment, building, handicrafts and skills, trading and other". Women's responsibilities made up two-thirds of "total labour utilized", men enjoyed two and one-quarter times more leisure time. Besides their agricultural work, women were responsible for close to ninety percent of domestic labour (Tibaijuka, 1984, p.77). Women, cleaned the houses, collected most of the firewood (a job shared by children) drew and carried water supplies from the nearest spring or stream. I followed four of the women in our sample to the nearest water supply. In all four cases the return trip took over one hour. The women carried a twenty five litre gerry can which she refilled twice per day. This means that each woman spent at least two hours per day drawing water, and, about one hour per day gathering wood. Women could not afford kerosene or stoves and therefore all sample households, including the richer farmers cooked their meals on three stones over open fires.

The Female Out-Migrant Returnee Household

Wives do more farm and domestic work than husbands but do not get a proportional share of the proceeds. After divorce or death of a husband a woman's situation usually deteriorates because she is unlikely to own land or other productive property. (Inheritance and control over land is discussed in the next section.) Aspiring women farmers must migrate to the towns in order to accumulate capital to buy and develop land; then they can eke out an independent living, but most make do with far fewer resources than male counterparts. Independent women farmers follow the classic model proposed by Ester Boserup: they adopt innovations in order to survive. Most women who head households in Kagera region are widowed or divorced and have lived outside their villages in order to accumulate capital, they are not women with absent migrant husbands, as was the case in Staudt's (1982) study in neighbouring Western Kenya. Swantz (1985, p.72) citing the 1978 Tanzanian census, reports one of the country's highest divorce rates (double that in Kilimanjaro, the nation's other densely populated banana/coffee growing zone). Though male and female-headed households grow identical staple food and cash crops (plantain bananas and coffee), and for the most part employ identical methods of hand hoe cultivation, they differ in many ways.

Aggregate data provides a distorted picture of material conditions in the villages. I found that gender differences were so great that it made little sense to talk about the "average villager". Households headed by divorced or widowed women display characteristics distinct from those of male-headed households, whether poor, middle or rich. Since female headed households are becoming more common in Sub-Saharan Africa, studies that fail to dis-aggregate male-headed and female-headed households risk misrepresenting both. When we consider the gender-specific traits of Haya households, a clear pattern emerges.

Female-headed households have smaller plots and insecure tenure, very few can hire regular or even part-time labour, or can afford agricultural inputs, most are denied access to extension services, they have far fewer opportunities to generate capital through non-farm employment and their education levels are much lower. Therefore the *economic imperatives that force men to migrate are even more imperative for women.*

Independent women farmers, practice a more labour-intensive and subsistence-oriented farming supplemented by petty trading activities and the production and sale of beer and distilled liquor. Female household heads are increasing production of beer bananas in order to generate a constant cash flow for essential goods and services, including food supplements such as maize flour. Women concentrate on beer production because it is a ready source of cash, however in the longer term beer bananas take land away from *food crops* and may threaten food security. Female household heads should be recognized as having not only a distinctive profile as smallholder farmers, but as having distinctive needs in terms of future development of their capacity for food production and income generation. They experience distinct social structural pressures that necessitate or at least encourage migration/return.

Case Studies

Before discussing the structures of gender inequality and how these impact on migrant/ returnees, let us turn to a few mini biographies, stories from the lives of women that highlight some salient economic and social relations.

Constance, is a local police woman, her case illustrates some reasons why women household heads want to be independent of men. She is already caught up in the out migration/ return cycle, she plans to have children with her male partner, nevertheless, she wants to avoid the constraints of official marriage and the bride price system. Her ambition is to save up enough money to buy a small parcel of land and set up an independent farm. Even if she succeeds, Constance will have to grow enough food on a small plot of land to feed her family. Unlike a typical male who inherits land, she will have less capital for farm inputs, trading or small scale industry since more of her accumulated capital will have to be used to pay for *land.* However were she to marry she would not be entitled to own land even if she paid for all or part of the plot. Any land initially belonging to her husband would revert to the clan should they separate or should he die. Furthermore, since her husband would probably establish a legal title to all common property she prefers to struggle and to remain unmarried.

The strongest role models for girls are the women who not only become independent but produce a surplus as well. In order to do this they must accumulate capital from a period of urban migration *and* then invest it in a sustainable revenue generating enterprise. They must effectively manage their primary enterprise as well as some additional sources of revenue that provide insurance against hard times or natural catastrophes. Consider the following biographies, Subira, Neema, Susanna, Maria, Anna, and Zipporah.

These successful role models followed two career patterns: three finished four years of high school (Form Four Leavers) and then went on to jobs, income, savings and accumulating capital. The other three had only elementary school education but accumulated capital through trading and invested it in highly profitable but also high-risk illicit enterprises.

Subira came to work in Bukoba after completing High School. She was hired as an accounts clerk but later became the common-law wife of her employer, an elderly European businessman. She inherited her capital from her former consort, who left her a plot, an improved cement house, a cow shed and seven purebred dairy cows; this was the source of her start-up capital as a successful independent woman. She rented out rooms in her house and sold milk: the dairy operation provided the lion's share of her income (about one thousand dollars per year). In 1994, when I last saw Subira, she was trying to diversify her interests by taking a training course to become a magistrate.

Neema was another out migrant who completed High School and then agricultural college. To acquire start-up and operating capital for a profitable sideline, she worked as a *bibi shamba* (an agricultural extension officer). Profits from her successful catering business supplemented her six hundred dollar per year salary. In fact, she derived the major share of her income (fourteen hundred and fifty dollars per year) from catering. She built this business up over ten years but its success was based mainly on her hard work. She woke up at four AM to prepare the fried cakes (mandazi) and meat squares (samosa) before leaving for her job. Her houseworker, a fifteen year old girl from an outlying district, delivered cakes and samosas to local restaurants, and then spent the rest of the day selling the remaining stock from a stand near the house. Although this extension worker's income depends largely on hard work, her salary and her initial savings were possible because her family had been able to send her to Secondary School: her education is a type of inheritance.

Susanna, is a former commercial sex worker who resided for six years on the Kenya coast an important tourism centre in East Africa. After saving enough capital, she returned to buy a small farm near Bukoba. She derived a cash income of about three hundred dollars per year from coffee and poultry production, but the largest part of her income was from making and selling yogurt, which brought her a profit of about eight hundred dollars per year. She also owned and hired out a bicycle that provided another one hundred dollars per year, she also received a two to three hundred dollar remittance from her daughter in Dar es Salaam in return for taking care of three grandchildren.

Maria completed primary school and obtained employment in Bukoba Town. Once she accumulated sufficient start-up capital she quit her job; she soon realized she could earn more money in business. She began as a seamstress (using her own sewing machine) and supplemented sewing revenue by brewing and selling maize meal beer. She used her profits to purchase a farm on a ridge overlooking Bukoba. She claims to have discovered her husband committing adultery and then divorced him. After the divorce she launched into a highly profitable but risky trade. She purchased a motorbike and began buying black market gold nuggets from the mines near Geita and reselling them to goldsmiths in Mwanza. This trade had been her major source of income, about three thousand dollars per year, and she wisely invested her disposable surplus in a second farm in the Bukoba area. Since a Canadian company, Sutton Industries, began a multi-million dollar commercial gold mining operation, in 1996, the bottom has fallen out of the black market in gold.

Anna was formerly a commercial sex worker (some in the community claimed she was still involved in the trade). She ran an illegal bar in a rented cement-walled house in the Bukoba area. She started her money-making career as a market trader, buying wholesale quantities of tomatoes and reselling them in small portions. She also grew and sold her own maize, potatoes and yams. She gave up vegetable trading when she found she could earn greater profits selling illegal home-brewed banana liquor. She combined catering (fried fish) with her bar operation; this proved a clever way to draw in more customers. The bar provided her with about fifteen hundred dollars per year in cash income. As an only child, she had inherited a farm in an outlying village where she was building a big cement house.

Zipporah's case was slightly different since she did not plan to return to the region, however, the basic need to accumulate capital during a period of migration was the same. Like Subira she married a wealthy businessman and became a widow in the mid 1980s but unlike the others she has not resettled in her natal region, nor have any of her six children. Her two sons are in University in India, two of the four daughters are married in Dar es Salaam and live close to their mother who owns a flat in a fashionable neighbourhood. One of the two younger daughters is a fashion designer in Nigeria, and the youngest daughter is a hairdressing student at a college in Nairobi, Kenya. Zipporah supports her family by renting out seven houses in the Dar es Salaam residential areas and imports and resells used and reconditioned cars from Japan, United Arab Emirates and Europe. Her family and import business keep her tied to Dar es Salaam but there is another reason she does not return to a village. After her husband's death she was living on his large coffee estate west of Bukoba town. Her husband's two surviving brothers claimed that the farm was their property as custodians of the deceased husband's clan. This claim was upheld in High Court and Zipporah and her family were evicted. Because of this decision, as well as her successful business interests, she does not want to return like the sampled women migrants residing in Kagera region.

Unlike areas of West Africa or the Caribbean, Kagera has fewer successful women traders and therefore few female networks of mutual assistance finance and technical aid. Women headed households comprised about one-third of the sample but only controlled about one-tenth of trading incomes.

Table 1:
Selected Variables in the Comparison of Male and Female-Headed Households

ITEM	MALE-HEADED (n=174)	FEMALE-HEADED (n=76)
1. Age of household head (mean)	54 years	44 years
2. Number of permanent household residents	5 persons	3 persons
3. Educational level of household head	5-8 years primary school	no formal education
4. Reported acreage per household	5 acres	1 acre
5. Number of coffee trees per household	585 trees	176 trees

6. Estimated banana harvest (mean number of bunches per year per household)	228 bunches	184 bunches
7. cows per household (mean)	3.9 cows	0.1 cows
8. Number of goats per household (mean)	3.5 goats	0.9 goats
9. Percentage of households with income from non-farm salaried employment	37%	7%
10. Percentage receiving agricultural extension services	46%	7%

Land and Power: The Structural Basis of Inequality

Guyer (1986,p.403) notes with reference to Africa as a whole, that "while women provide the main agricultural labour force, their rights to the land are insecure". Certainly in the Kagera region of Tanzania, as Swantz (1985) and Bader (1975) have asserted, customary land tenure arrangements are the primary factor perpetuating the unequal position of both married and unmarried women farmers. Though the Villages and Ujamaa Act (1975 revised in 1983) provided for democratic allocation of land to all village members, the de facto situation in Kagera Region was that it had fewer official *Ujamaa* villages than any other region and the villagization process did not displace customary land tenure (Bader, 1975; Boesen, Madsen and Moody, 1977; Raikes, 1978). By the 1980s, less than three percent of Kagera's villages were registered as Ujamaa (McCall, 1987, p.194).

Control over adequate land is the basis for food security and for the accumulation of capital through farm expansion and cash crop production. Female household heads, because of land deprivation, are able to secure neither a truly reliable food supply nor the expansion of their productive capacity. The female-headed households in this sample had far less land at their disposal; their reported average holding was two acres versus four acres for male-headed households.

There is little or no opportunity for women to acquire ownership of land under customary law (Swantz, 1985, pp.67-68), since even in the absence of a direct male heir, the land will remain in the control of male relatives rather than passing to a wife or daughter.[3] As a result, many contemporary female household heads are merely *custodians of clan land* and will never be granted a title deed; they must live in apprehension of eviction by male relatives or in-laws. The other avenue to land ownership is the purchase of alienated land. In 1918, the British established a system by which land could be passed on through customary law or could be converted to estate property with no customary restrictions (See Reining, 1970). The government, elected in 1995, is enthusiastic about continuing the trend toward economic liberalization and streamlining land sales and transfers, but in densely populated Kagera region, land held under clearly-registered easily-transferable title deed is very scarce and therefore very expensive. Generally, land is available for sale only when the owner is deeply indebted or has no relatives who will inherit the land; as Weiss (1993, p.30) puts it "land that is inherited . . . is a semi-alienable property". Extended family members can and frequently do contest land sales in rural areas even if the land has a registered title deed. Banks in East Africa generally will not accept land outside municipalities as security on loans.

Only seventeen percent of the female interviewees had purchased all or part of their holdings, in most cases from relatives. In contrast, eighty-five percent of the male household heads in the sample owned their principal plot and thirty-four percent of them had added to their land holdings through purchase. The majority of women household heads occupied land under the ownership and jurisdiction of male relatives, most often without any assurance that the house and land would not be taken over at any time by a male heir. (See table 2) Land tenure conditions for female-headed households are thus insecure and not conducive to food and cash crop production on an equal basis with male-headed households.

Table 2:
Source and Type of Land Tenure for Female Headed Households (N=76)

Source of Land	Type of Tenure
Landless	24%
Purchased title deed	13%
Inherited occupancy from husband	28%
Inherited occupancy from father	17%
Purchase, inheritance	4%
Ujamaa tenure	14%

Is Bride Price a Patriarchal Custom?

Male small holders are privileged members of their patriarchal culture: they usually inherit land and they benefit from the hard work and resourcefulness of wives and mothers who have worked to develop that land. Out migrating males exploit women's labour to develop family plots and to take care of them while they are away earning money. Marriage norms require payment of bride price and once it has been paid a wife and her family are locked into a restrictive set of social obligations. In the past, the groom or his family donated a cow to the wife's family: they also provided banana beer and meat for the wedding feast which usually lasted two or three days (Cory and Hartnoll, 1973). Since the 1960s the Haya have changed these customs: bride price is paid in cash and costs about a year's annual income, worth anywhere from four to twelve hundred US dollars in the 1990s.

The bride becomes the primary domestic worker and child bearer; a husband expects a wife to be submissive and obedient. He makes the reproductive decisions. Should the woman be childless that is grounds for divorce: a husband can also divorce a wife who drinks at neighbours houses or in bars since this presumes drunkenness and low morals as does travel outside the village even if ostensibly for business or social reasons. Similarly, if the husband finds his wife lazy or if she fails to satisfy him sexually this is also grounds for divorce. As Weiss (1993) notes, senior women from both her own and her husband's agnatic clan, will be harsh with new wives; should she not satisfy her husband this is taken as evidence or poor training and socialization. After a divorce or should a wife run away, *her* family is liable to pay back the bride price: the patriarchal pattern continues because both families back the marriage contract. Although the Haya are not a "father right" society and generally a woman or her family of origin retains custody of her children. White (1990) emphasized that Haya prostitutes sent money back to parents in order for them to repay bride price obligations. Despite the risks of HIV infections — double for divorced or separated partners— marital instability remains a pervasive social problem one that has not diminished even with increasing AIDS awareness (Nabaity, et al., 1994).

The norm for wives and mothers is to enhance the success of the male-headed household. Women either work on larger farms which they usually cannot inherit or they remain unmarried or divorced and eke out a living any way they can. Independent women are forced by necessity to occupy small plots and work intensively to survive. Because of the bride price system, a woman seeking a divorce cannot usually expect help from her family of origin, because her family fears being forced to pay compensation to the ex-husband's family.

Food Production in Female-Headed Households

A woman who achieves her objective and acquires land, faces another set of challenges. The land tenure system forces female-headed households into labour-intensive food production — more intensive, in fact, than cultivation by women farmers in male-headed households, because wives have additional responsibilities (weeding and harvesting) associated with their husbands' coffee trees.

All households studied grow the staple crops of plantain bananas and beans (the latter inter-cropped by women under the banana trees). In addition, women in all female headed households and one-half the male headed households cultivate *omusiri,* supplementary crops grown on communal land outside the banana and coffee plantation: principally maize, cassava, sweet potatoes, and ground nuts. The range of crops grown is identical, yet female-headed households generally cultivate bananas of all types (plantain, sweet, and beer bananas) more intensively, though their overall production is lower. The average male-headed household has one acre of cultivated land per person and the female-headed household has one-third of an acre, both produce an average eighty to one hundred banana bunches per acre, but the independent women farmers do it lacking access to inputs (such as hired or wife labour, manure, and agricultural extension services). Independent women farmers are evidently concentrating their only available resource— household labour — on production of the staple food crop; thus they demonstrate the potential for increased food production even on small plots.

Cooperative Work Groups

Out of necessity, women have developed ways to bolster farm production by forming cooperative associations or the *kyama* — these associations are modeled on the traditional male cattle owners guilds described by Cory and Hartnoll (1971). The *kyama*, described by Swantz (1985) and Boesen (1973) is an association of women who cooperate as field workers and provide a social support network:

> The women's *kyama* members ... spent their time in turn on each other's fields, doing the complete cultivation of one woman's field in one day. The harvesting was done in the same manner. They also helped each other in family celebrations and gave special contributions of money and work when any member was ill or in special need, or when a family member had died. (Swantz, 1985, p.63)

The women's *kyama* is the only viable cooperative work group I observed anywhere in the region.

Four-fifths of the 189 women interviewed belonged to a *kyama*.

Women who had enough land to produce a surplus of beans (and occasionally other crops) would sell it and share the proceeds. Those such as the wives of richer farmers, who contributed more land to the *kyama* got a larger share of both produce and cash (personal communication Margaret Bishubi). We observed a tendency for women to form *kyama* only with members of the same congregation or at least the same religion; this verifies the findings of another researcher. (Bader, 1975).

Cash-Crop (Coffee) Production in Female-Headed Households

Coffee is the longstanding regional cash crop: during the boom periods of the 1920s and the 1950s, Haya farmers planted millions of coffee trees— they grew coffee to provide money for housing, school fees and farm improvement. Despite bad harvests in Brazil and record prices in the 1990s the present system does not favor coffee production, especially for women farmers.

Women are effectively excluded from extensive coffee cultivation by its customary association with men, and the small size of women's farms relative to the greater number of persons they support per acre. On average, the female-headed households had only one-third the number of coffee trees (table 1, item five). Only eighteen percent of female-headed households were planting new coffee trees to expand production, compared to the forty-two percent of male-headed households. Planting coffee seedlings and tending coffee trees is traditionally regarded as men's work, and women have usually been excluded from training and hands-on experience in important skills such as grafting, pruning, preparing seed beds and correct trans-planting; they are also excluded from coffee profits. Women carry out the tedious jobs of weeding and harvesting but lack experience in the crucial tasks related to expanding coffee cultivation. This has been an enormous disadvantage for women in the mid 1990s as world coffee prices reached record highs.

Divorced female household heads (two-thirds of the total) are also deterred from coffee farming if they are occupying land owned by the ex-husband or his clan. Ex-husbands or their male relatives can and do exercise the right to claim the proceeds of the coffee harvest, leaving the woman farmer with little incentive to work on the coffee crop.

Since independent female farmers use their available time to produce the "women's" food crops, they must obtain hired or voluntary help (from male relatives) if they want to cultivate coffee. Sixty percent of women with coffee trees relied on occasional male assistance to prune them. However, only thirty percent of female household heads were able to afford even occasional help and thus most were not consistently planting, pruning, weeding or harvesting coffee. In contrast, sixty percent of male household heads used occasional workers and the wealthiest hired permanent workers for coffee-related tasks.

Differential Access to Hired Labour, Agricultural Inputs and Agricultural Extension Services

Female migrant/returnees have little or no access to the major inputs (wife-labour, regular or permanent workers, cow manure as fertilizer, and agricultural extension services) used by male farmers to intensify cash crop production. Wives are responsible for the lion's share of food production — beans, maize, cassava and sweet potatoes, as well as much of the banana cultivation.

Fertilizer, cow manure, is also unavailable to most women farmers — only one-third of woman interviewed could obtain it. This is a significant handicap, since the addition of sufficient manure can triple tree crop yields of coffee and bananas, which is why three-fifths of male farmers use it. Though few households in the sample (fifteen percent) used insecticide and sprayers, none of those who did were female-headed.

Indeed, the use of insecticides, improved farming techniques and even proper pruning require expert assistance and instruction. Female household heads are dis-advantaged from the start by their lack of primary education, and thus their poorer grasp of Swahili, the national language of education and government communications. Two-thirds of female household heads received less than two years of formal education. More women household heads than men attended adult education classes, but none of the women's programs featured agricultural education.

Very few independent women farmers (7% see table 1) are able to make use of the agricultural extension services available in their village. These services are provided by on-farm demonstrations and assistance (for example, in pruning techniques or in the use of green and animal manures), or at courses given in the nearest training institute. Thirty-three percent of male-headed households received either on-farm assistance or attended a course; only nineteen percent of woman reported receiving extension assistance. Staudt (1982), highlights exactly the same problems faced by women in Western Kenya; Haya: women farmers are excluded from informal but vital networks and venues of communication where agricultural information is exchanged among men. Men frequently go out in the evenings to drink home-brewed beer and illicit moonshine, they socialize and exchange agricultural and other information, while their wives and women heads of household stay home with the children.

Access to Capital: the Pattern of Income-generation and Off-farm Employment for Men and Women

The migration cycle is the principal route to accumulation of capital from trading or salaried employment and subsequent success as a village smallholder. Successful investment of capital was the key factor in determining who became and remained a rich farmer. However, capital accumulation varies greatly for men and women. Sixty percent of the male household heads surveyed used capital from salaried employment and trade to expand or improve their farms, contrasted to twenty-two percent for women. Men are less often *compelled* to migrate but they tend to derive greater benefits if they do so.

Male farmers were able to undertake protracted absences from the village, of up to twenty years, because wives and children operated their farms. They frequently returned with sufficient capital to purchase land additional to the inherited plot, to hire permanent labourers, to build improved houses, to buy cattle, and, occasionally, to invest in transport vehicles or other non-farm business ventures which provided additional capital for farm expansion.

The most successful Haya farms have always been integrated enterprises consisting of both farm and capitalized business activities such as shops, transport, watch repair, tailoring, butcheries, and furniture-making. Women farmers specialize in craft industries which require very small inputs of capital; such as, brewing, distilling, collecting herbal medicines for sale, midwifery, weaving mats and baskets. Women have less time to devote to these trades and, with the exception of traditional medicine, do not need skills considered specialized by Haya culture. As a consequence, the male farmers interviewed reported an annual (mean) cash income of $760 (US Dollar equivalent in Tanzanian shillings) compared to the US$370 (TSH equivalent) average income reported by female-headed households.[4]

Migrant Women, Wealth and Commercial Sex Work

Haya women are notorious as commercial sex workers in the major East African cities, such as Dar es Salaam, Mwanza, Kampala, Nairobi and Mombasa (Hyden, 1969; Southall and Gutkind, 1957; Stevens, 1995; Swantz, 1985; Weiss, 1993; White, 1990). In colonial times Nairobi-based Haya women became notorious because of a very visible minority of practising commercial sex workers. Luise White (1990) in her impressive study of the history of prostitution in the regional metropole, depicts Haya women as the pioneers of a new more aggressive type of prostitution that entirely changed the trade. Previously, sex workers rented huts in Pumwani (a Nairobi neighbourhood) then waited discretely for customers to find their homes. Haya women, in contrast, sat outside, called out prices and even banded together to beat up customers who would not pay. As White also points out, Haya sex workers did not buy houses in Nairobi, but returned home to set up independent households;

> In the time I interviewed in Pumwani I could not find one Haya woman who had been a prostitute between 1936 and 1946. All had gone home. (White, 1990, p.110)

Because they had money to buy farms, returned sex workers gained status and became female role models. Bader (1975) Stevens (1995) and Swantz (1986) all present cases of Haya women who returned to purchase farms with capital saved from urban prostitution. As Bader (1975) points out, many women out-migrants who return to set up farms were classified by their village neighbours as returned commercial sex workers even if they had been white or blue collar workers or business people. Brad Weiss (1993) collected 257 marital histories. He reported that twenty-two percent of his female informants "spent some time in urban areas in activities associated with selling sexual services".

Prostitution has been and continues to be a significant means of capital accumulation and land acquisition for Haya women, especially the divorced. Swantz (1985,p.74) interviewed fifteen former commercial sex workers residing in Bukoba District villages in the 1970s (most of them divorced) and discovered they had spent an average of seven years in urban prostitution but then returned to the region and bought land and became independent small holders.

There were only four women from this sample who admitted to being returned commercial sex workers. Our data on out-migration from the 250 households indicates significant recent female migration to take up white-collar employment. Interviewed parents reported that fifty-eight percent of their female migrant children were students or employed in white/ pink collar jobs such as government employees, teachers, nurses or secretaries. However as Weiss (1993) highlights "dislocation" and "uncontrolled mobility" are the "quintessential acts of cultural disruption" not commercialization of sex.

Only a very special minority of women can take advantage of the existing opportunities and use out migration as a means to become rich. Paradoxically, successful independent women are more likely to have been brought up in male-headed households who could send children (especially sons) to private secondary school. In the sample there were one hundred and seven secondary students, of whom eighty percent were males: moreover only ten percent of high school students came from women-headed families. Female headed households were more egalitarian; although proportionately fewer children could be sent to high school, there were equal numbers of boys and girls. Since women have less education, fewer can enter the civil service or other relatively secure forms of paid employment.

Women who transcend patriarchal restrictions need drive intelligence and good luck. Gender plays a major role in acquiring riches. Only five of the sixty sampled households headed by women had high enough incomes to be considered rich, even by local standards. If gender were a neutral factor, we would expect there to be an equal proportion of rich women to rich men: there should be seventeen. The cases were already presented as mini biographies (Zipporah as a resident of Dar es Salaam was not part of the sample). These stories illustrate how few women were able to overcome gender as an economic handicap.

The "rich" women had important characteristics in common: all were younger than their male counterparts (four were in their thirties and one in her early forties); all lived in close proximity to Bukoba town where there were more numerous and potentially more profitable opportunities for enterprise. All were independent of men — a widow, three divorced, and one who never married. (None of the 113 wives I interviewed had personal incomes which would put them near the $1250 US yearly cash minimum used to demarcate "rich farmers".) In the reported case-studies, the basis for their primary accumulation of capital was either inheritance or illegal activities. Two of the women accumulated their initial capital from prostitution and one from an illegal bar, and the two widows inherited their capital goods and houses from older husbands. This indicates that women who follow the culturally prescribed system of marriage and child-rearing cannot expect to accumulate wealth. (Both widows, Subira and Zipporah, had been much younger *second* wives of rich men). Women who accumulate capital, must sooner or later become independent of men. These richer women had to be more hard-working, innovative, and efficient at organizing their affairs and protecting their property than their male counterparts.

Beer and Liquor Production in the Economy of Male and Female-Headed Households

In the sample as a whole, brewing makes up one ninth of aggregate income but in the female out migrant/returnee household beer related income accounts for about two-fifths of the total. (Coffee accounts for 14 percent and salaried employment 12 percent and small trade and crafts for 54 percent). Beer and liquor are sold commercially in innumerable household sitting rooms and at kiosks scattered around the villages, each of which attracts a (mainly male) clientele of twenty to thirty persons per evening.

At each stage of the operation, beginning with beer bananas and ending up with commercial sale of liquor, value-added profit increases. Most households in the villages engage in some facet of the beer banana enterprise. The data (household interviews, informal participant observation, interviews of bar owners and workers) indicates a pattern of gender inequality, or to use McCall's (1987) terminology "male hijacking" of the illegal and more dangerous but more profitable large-scale public sale of liquor. Males dominate the "*konyagi* revolution" (Tibaijuka, 1984) the use of petrol drums as boilers and submerged pots as collectors. The drum method quadruples production, but it fit more clearly into a male activity pattern because of the need to establish hidden stills near streams and to evade and bribe the police and local vigilantes (the *sungu sungu).*

A good illustration of why males predominate in larger-scale bootlegging, is the case of a large operator who was arrested in the village of Muhutwe (on the main trunk road) in October 1990. His Land Rover contained one thousand litres (fifty gerry cans) of banana liquor. A local official told me that this bootlegger was under surveillance and known to local politicians who wanted him arrested. The same source believed the bootlegger bribed the chemists who analysed the liquor and the magistrate who presided over the case: he did have to pay a fine but evaded a prison sentence. In the present social context this type of wheeling and dealing is usually a "man's world" open to only the most exceptional of women.

Nevertheless, small-scale production and sale of beer and liquor is the single most important source of cash for women. In the five sampled villages, two thirds of women farmers reported selling beer or liquor: in one village, all but one woman derived the largest share of her income from this source. This finding agrees with another Tanzanian study (Cecelski, 1984) who reports that fifty-five percent of women claimed brewing as their main occupation.

As with the food-banana crop, male-headed households produce a larger overall volume. But independent women farmers produce more beer from less land: surveyed female-headed households produce about 300 litres for every acre of land, while male-headed households produce 200 litres.

Patterns of Expenditure in Male and Female-Headed Households

The female-headed households in our sample were dis-advantaged materially as a result of a lower cash income from fewer sources, and they were also under more pressure to use the available cash for foods and essential services.

The female-headed households were less likely to have improved houses made of fired brick or cement instead of sun-dried mud—one-third of women's houses compared to one-half of the men's houses. A significant number of female-headed households were without any furniture except mats (fourteen percent) or had only a single bed (twenty-one percent). Two-thirds of female-headed households had fewer consumer items than the median rank for the sample.

No female-headed household had a car, only eleven percent possessed a bicycle (all were inherited rather than purchased new). Bicycles were rented out to men who used them to transport crops or to reach a place of employment. This corresponds to the pattern noted by Bryceson (1990) and Mbilinyi (1978) that women's income is spent on household reproduction needs and very few women can accumulate savings for luxury consumption.

A comparison of regular household cash expenditures revealed that male and female households spent cash income on certain identical items: but the composition of this expenditure is significantly different. The female-headed households were using one-third of their entire available cash income to feed themselves and family, giving special priority to improving the diet of their children. Male-headed households spent one-sixth of cash income on food. Moreover these figures can be misleading since wives often use their own cash incomes to buy special foods for their children. Children were more likely to get breakfast (maize porridge, milk and sugar) in female-headed households than male-headed households despite their lower income: sixty percent of female household heads compared to fifty-one percent of male household heads, reported their children ate breakfast regularly. Guyer (1986, p.401) describes a parallel phenomenon in Cameroon: "Women (wives) produce most of the staple food and spend up to three-quarters of their income on food supplements, and household needs".

A relatively well-to-do female household head in the ridge-top village sub-sample, provides a good illustration of this process. This village teacher reported an annual cash income of $700 TSH equivalent. Unlike many male-headed households who produce all their own food and use wives and older children for farm labour, she was spending cash on food, and salaries of young live-in labourers hired to keep up subsistence food production for a household of ten persons residing on a 1.5 acre plot. The teacher's household budget illustrates some of the constraints faced even by women with salaried employment. Out of her cash income (about twice the average for the female-headed households in the sample), she was spending seventy-six percent on food, nine percent on school and hospital fees and ritual or social obligations, five percent on taxes, and the remainder on other household needs such as clothing and building materials. At least one-quarter of the male-headed households in our survey, on the other hand, had incomes high enough to buy the essentials and retain a surplus. That surplus could be invested in home and farm improvement.

The cash flow problem of female headed households is aggravated in times of severe food shortage due to banana crop failure. Four-fifths of household heads interviewed recalled at least one *wakati wa njaa* (season of hunger) during the preceding decade. Most frequently mentioned were the country-wide shortages of the mid-1970s, the 1979 war with Uganda, and banana crop shortfalls in 1982, 1985, and 1986. At each period, many households were forced to buy maize flour and cassava at black market prices. The families who did not report any food shortages were all male-headed, operated larger-than-average plots and reported cash incomes over $600/year, TSH equivalent. For the average female-headed household, cash expenditure on food supplements (including staples, like processed maize flour) is becoming a chronic necessity.[5] The situation may be aggravated by the conversion from food banana to beer banana cultivation — the only available income generating activity which is both culturally acceptable for women and brings a fast and constant cash return.

Conclusion

This study suggests the need for further investigation of female-headed smallholder households and the dynamics of migration as a survival mechanism. Women heads of household have distinctive characteristics which dramatically affect their contribution to aggregate food and cash crop production. In many ways their profile (of land shortage and reduced access to inputs and capital) resembles that of poor male-headed small-holder households. It should be emphasized, however, that the female household heads in Kagera, unlike the poor males, faced constraints that necessitated migration and in some cases encouraged them to engage in commercial sex work— despite the risk of HIV infection. Their children will face the same constraints and be forced to migrate in perhaps even greater numbers. Young Haya women lack access to inherited land under customary law, they are educationally disadvantaged and they are deterred by customary norms that exclude them from serious involvement in coffee production. Marriage is no guarantee of material security and may often make life even more difficult for women. Women who out migrate and return need improved access to land and inputs as well as income generating alternatives to the beer trade, which though it provides for immediate needs could ultimately jeopardize food security for many households.

It is encouraging that women farmers, despite economic disadvantages perpetuated by a patriarchal culture, support themselves by being well organized and hard working. They set an example of enterprise and reliability. During part of my fieldwork in the "ridge top village" I stayed in a mission hospital guest house. Mary, who lived next door, was an obstetrics nurse with seven children of her own. She put in long hours to do her job at the hospital and care for her farm in the adjacent village. Her thirteen year old daughter took charge of a one year old son. Mary tended a garden at her hospital residence as well as her farm six kilometres away. She was determined not to succumb to the injustice of the landholding and bride price system and to use the traditional female skill midwifery in the more modern hospital setting.

Notes

1. Some scholars argue that sub-Saharan Africa although predominantly patrilineal was not patriarchal, at least not in the sense of society exerting strict control over women's sexuality. However those who reject the idea that African men generally exert control over women's sexuality eg. (Caldwell, Caldwell and Quiggin, 1989) define the contrasting "Eurasian" pattern in terms of "an *obsession* with controlling women's morals and mobility" and not the *ability* to exert control. As Weiss (1993) points out, women's mobility equates with evasion of the Haya's androcentric, agnatic clan centred patriarchal norms and in fact he shows how the clan leaders are *obsessed* with a situation that they criticize but cannot control.
2. Our operational definition of the household is: a group of people who normally eat and sleep on the same plot year round. The sample was arrived at by choosing one household from each of the "cells" of ten households organized by the then ruling party (Chama Cha Mapinduzi). Since each cell of ten households is a geographical cluster, this method ensured coverage of all areas of the village. We interviewed male and female household heads (and the spouses of the former), and in addition carried out a census of household members and migrants. Interviews were carried out in Swahili, using an interpreter in cases where the interviewer preferred to use Luhaya; the average interview lasted two hours, though we made repeat informal visits to many households.
3. Swantz (1985) cites the case of a revivalist Christian in Bukoba

who willed his land to his wife and a divorced daughter in the absence of a male heir. Title was granted by the courts to the deceased's brother (who contested the will), though the women were allowed occupancy rights.

4. We arrived at estimations of income by extrapolating from known quantities. For example, coffee incomes were calculated based on the number of kilos sold to the Cooperative and the fixed price cited by farmers and the Cooperative. For income from the sale of banana beer, bananas, food crops, dairy products, livestock, firewood, handicrafts, traditional medicines, etc., we carefully obtained average quantities sold per week, verified local market prices, and probed to discover seasonal or monthly variations. Salaries were always verified against government wage scales.
5. Unlike most areas of Tanzania (other than Kilimanjaro), Kagera is not a maize producing area and maize porridge (*ugali*) is not the traditional staple dish. Though households were producing some maize, it was not cultivated intensively, and most of the households studied were regularly buying supplies from the shops in processed form. The burden of such cash outlays, however, rests more heavily on female-headed households.

11 Consumer Goods and the Cash Economy

The Tanzanian national census statistics for the Kagera Region appear to confirm Goran Hyden's concept of the "peasant mode of production" and the "uncaptured peasantry".

There is a tiny official commercial infrastructure for a population of one million and the Haya appear to be subsistence farmers who sell coffee to earn cash and might withdraw from the market when prices are low. This is the type of local economy I expected to find when I first visited the region in 1987. However, since Independence there have been four clear changes in Haya consumer behaviour: 1) purchase of a greater quantity and variety of local foods; 2) increased reliance on sugar and flour — mainly imported from outside the region; 3) purchase of a greater number of consumer goods as part of the acceptable standard of living; 4) decreased consumption of bottled beer and cigarettes and increased purchases of locally distilled liquor.

Before Independence most of the purchased articles were imported into the region by the Colonial government. In 1954, the consumption of merely eight imported commodities reached almost 2,000,000 pounds Sterling.

Table 1: Consumer Imports, 1954

cloth/piece goods	£500,000
corrugated iron	£360,000
cigarettes	£300,000
imported rations	£300,000
cattle for slaughter	£225,000
bicycles	£75,000
radios	£48,000

Source: Tanganyika Provincial Commissioners, 1954, p.61

Today the Haya purchase a wider range of commodities, many of which are locally produced. Twenty eight household outside the sample but residing in the same villages, agreed to supply me with a daily record of all their cash purchases during the first five months of 1989 (see the appendix for the complete list).

These families regularly purchased thirty six items, see table below.

Table 2: Average (Mean) Monthly Expenditure (in US Dollar Equivalent Terms) by Item, in Descending Order

clothes	3.48	medicines & hospital fees	0.20
sugar	2.21	salt	0.20
fish	2.09	soft drinks	0.18
banana beer and liquor	1.01	snacks	0.17
meat	0.90	karoboi	0.15
flour	0.83	buckets & wash basins	0.14
rice	0.69	batteries	0.14
soap	0..67	stationery	0.13
kerosene	0.64	tea	0.13
kitchen wares	0.64	milk	0.12
bread	0.58	cosmetics	0.11

bananas	0.57	cigarettes & tobacco	0.10
shoes	0.47	cooking oil	0.45
tubers	0.09	sheets, blankets, towels	0.31
beans	0.07	matches	0.07
pots and pans	0.30	needles & thread	0.04
fire wood and charcoal	0.25	fruit	0.04
vegetables	0.24	spices	0.02

The Commoditization of Food and Everyday Needs

These purchases signal a shift away from the subsistence agriculture of the 1950s and early 1960s, when bananas, beans and fish (the main ingredients in *matoke*, the basic Haya meal) were only purchased during times of famine. 'Luxury foods' such as flour, sugar and rice would be purchased only at holiday times, see (Ngaiza, 1980). Food now accounts for eighteen of the thirty-six items in Table one — about one-half of expenditures. Most of this expenditure is on staples— plantain bananas, fish, sugar, flour, cassava and other root crops, cooking oil, salt, milk, and beans. In the 1950s (according to five elderly key informants) there was little use of sugar, flour, and tubers. In the 1990s even the poorest households must use their meagre cash reserves to buy expensive substitutes for their inadequate subsistence production. This trend indicates a decline in nutrition (along with food security) as cassava and processed sugar replace plantain.

Sugar, is a commonly purchased item. In the 1950s list, sugar imports ranked below cigarettes: in table one, sugar ranks second and cigarettes thirtieth. The Haya also purchase prepared foods made with sugar, such as cakes, prepared snacks and soft drinks, as an occasional treat. It appears that milk consumption has declined with the degradation of pasture land and reduction of the local herds and access to cattle (see Kjekshus, 1977; Ndagala and Anacleti, 1980). Since milk is generally served to young children, loss of pasture indicates a probable decline in child nutrition because few Haya can afford to purchase milk.

The pattern of purchasing firewood and charcoal also indicates a shift away from a subsistence economy and a possible decline in the standard of living. Each year the Haya spend more and more on the purchase of fuel: this significant transformation indicates how continued population growth and commercialization of resources have jeopardized fuel self-sufficiency. Until very recently the family plot and communal forests adequately supplied domestic needs. Because of the declining size of plots and communal forests, twenty out of the twenty eight households now purchase firewood on a regular basis. Some enterprising rich and middle farmers are taking up tree farming. Eucalyptus trees are the crop of choice because of their rapid growth: they produce a cash income after five years, as quickly as coffee. Firewood is always in demand because it is the only fuel used for cooking. None of my sampled farmers were using a simple cheap and efficient method of saving fuel, clay ovens or burnt-clay enclosed firepits. However, these innovations are likely to become popular in the near future as fuel becomes scarcer and more expensive.

Kerosene stoves are available as another alternative to the traditional three stone cooking fire, but they are too expensive. In two years of field research I did not see one villager cooking on a kerosene stove. Kerosene ranked seventeenth in the goods-purchased list, it is used primarily for lamps. The entire 250 household sample possessed a *karoboi* (a small lantern locally produced by soldering together cans and strips from cans). A few households also had a regular "storm lamp". Storm lamps are more durable and supply more light, but they cost forty times as much to buy, require more kerosene, and the glass covers and wicks must be replaced regularly.

The consumption patterns of the Haya corroborate the thesis that local initiatives work better than grandiose schemes. Two of the four most popular commodities, fish and alcoholized banana beverages, are produced locally on a small scale. Fishing requires little capital, since most fishermen use homemade outrigger canoes: sail boats are rare and small outboard motors extremely rare. Fish is sold from a large number of venues, and daily purchases can be made in almost every village within thirty kilometres of Lake Victoria. Pilot projects such as the Igabiro Fish Farm are a promising trend for the future as population increases, and the water hyacinth and human waste threaten the ecology of Lake Victoria and fish stocks. Tilapia, the preferred fish, can be raised almost anywhere including rice paddies. Although the Haya are used to fish from Lake Victoria and dislike the muddy taste of fish raised in ponds.

Clothes are the largest expenditure item in table one. Most of the clothes purchased in Tanzania, Uganda and Kenya are *mitumba*, used clothes or manufacturers remainders. These clothes are either sold by weight or donated by Western charities. According to a Canadian Broadcasting Corporation news report (Francophone service November 15, 1993) this trade generates an annual profit of about 24 million US dollars in Africa. A series of middlemen in Eastern Africa buy and resell them: importation to Tanzania, and distribution requires many intermediaries between initial purchase of the bales of clothes and a village buyer. Because the original price (determined by weight) is below any imaginable costs of production, used clothes are affordable even for poor villagers. Since it is easier to import used clothes into Tanzania than Kenya, another industry has developed, smuggling. *Mitumba* enters Kenya and Kenyan soaps, beer and other manufactured items move back into Tanzania. The official reason the Kenyan government restricts imports of used clothes is to protect the national clothing manufacturing industry, although it is rumoured that restrictions actually protect a few corrupt dealers with high level connections who avoid most tax and duty and are sheltered from competition.

Those who can afford to buy cloth, prefer to have local tailors make up their clothes. Local sewing machine owner/ operators occupy an important place in the informal sector. Tanzanian public schools require the students to wear uniforms, khaki shirts and shorts: these are worn out by students and are rarely available from second hand clothes hawkers; families must buy cloth and hire tailors. The school uniform requirement, provides an injection of capital for the local tailors, but this process is a financial burden for the poorest families and therefore a disincentive to education.

Mitumba (used clothes) are beneficial to local residents. Even though some tailoring income is lost, trading generates its own revenue and cheap clothes are an important incentive good. There are no viable alternatives to used clothes (Cookesy et al., 1986): without *mitumba* many small farmers and their families would have only rags to wear, as was the case when I lived in Tanzania in the late 1970s.

The sugar trade is a vestige of the 1970s and early 1980s official command economy and in 1990 prices were supposed to be state controlled. In reality, small farmers would buy sugar outside the nationally organized distribution system. Subsidized national sugar production is inefficient and expensive. Tanzanian sugar was produced at costs well above world market prices, but supplied to a regional trading monopoly at subsidized prices below the world market level. The Regional Trading Corporation (RTC) was supposed to supply the public but usually its supplies were monopolized by black market traders who payed out bribes in order to purchase sugar at subsidized prices. Retail sugar was sold on the parallel market by the village general stores, at prices usually about double the official level. With the election of a new government dedicated to trade liberalization this system has been deregulated and hopefully real prices at the village level will stay lower.

Services

The smallholder household needs a cash income for both goods and services, such as transportation, medical expenses, and social obligations. These expenses are listed below in descending order.

Table 3:
Average Annual Cash Expenditure on Services

Transportation	$8
Cash and gifts for relatives	$7.5

School fees and supplies	$6
Hospital fees and patent medicines	$5
Contributions to religious organizations	$2.5
Traditional doctors and medicines	$2
Funerals	$1

* All figures are medians based on conversion from Tanzanian shillings to equivalent in US dollars. N = 250

Social obligations (funerals and assistance to relatives) is the largest expenditure category, followed by transportation, medical expenses, education, and contribution to religious organizations. The high transportation expenditures are entirely consistent with the prevailing pattern of out migration. Given their limited cash flow, the Haya spend large sums visiting dispersed family members, however this is both a social obligation and an excuse to transport trade goods.

Education expenditure ranks below transport and family assistance for two reasons. The government provides highly subsidized primary education, and some state funded secondary schools; however many Haya are no longer willing or able to pay school fees to send their children to private or even much lower cost public schools. In the past, education was the key to social mobility — but by the mid 1980s the highest incomes were earned in trading and independent craft industries. The government sets unrealistically low salaries for posts, which require a high school certificate or even post secondary education. Because of the shortage of places in the low cost government-supported Secondary School system, parents of the majority (who do not qualify for state schools) must make a great sacrifice to pay the yearly fees of $200 required in the mission schools.

Unlike education, most transport expenses have an immediate payoff. So much of Haya income requires individuals and commodities to move from farm to market. Since the official bus services are infrequent and unreliable, most transport is carried out by unlicensed operators of *dalla dalla* (an adaptation of dollar dollar) — the Tanzanian equivalent of the Kenyan *matatu*. Owners of pickups and lorries, as well as company drivers, charge a reasonable fee to transport passengers crowded into the cargo area of their vehicles.

This expenditure data corroborates my argument that official state run institutions are less important to villagers than local initiatives. Even in the fields of modern education and health, private institutions are flourishing. In the Bukoba Rural and Urban Districts, which contain one-half of regional population, there are two Government High Schools but five Mission schools. Those who can afford hospital fees will spend hours travelling fifty kilometres out of town to the Catholic Hospital at Kagondo or the Lutheran Hospital at Ndolage for treatment of serious ailments or dental work. They choose tedious travel over easy access — there is a large regional Government hospital in Bukoba Town— because the standard of care and availability of medicines is superior. Informants spent four-fifths of their medical care budget at the mission or private hospitals.

Consumer Durables

The Haya purchase three categories of durable goods: 1) essentials, 2) middle range luxury goods, 3) elite articles. All but the very poorest buy category one goods, the essentials; two-thirds of households purchase the middle range articles.

Table 4: Durable Consumer Possessions: Percentages Based on 250 Household Sample

Essentials		Luxuries		Elite	
bed sheets, blankets and towels	99.4%	storm lamp	74%	motorcycle	8%
kitchenware	99%	radio	65%		
karoboi	98.8%				
beds	97%	bicycle	46%	car, pickup	7%
chairs	95%				
shoes for all in family	86%	cassette	34%		
table	85%				

Imported Mass Produced Goods

In table three, the "essentials" and "luxuries" categories are of two types: 1) mass produced manufactured goods, and 2) locally produced goods. Mass produced articles include: towels, blankets, bed sheets, plates/ bowls/cups (usually plastic), cutlery, cooking pots, buckets and gerry cans, most shoes, and all the goods listed as luxuries. Factory produced imports are purchased at the cheapest possible prices and are therefore of poor quality and need frequent replacement. Even the "elite" goods are purchased at the lowest possible costs: East Africa is a booming market for used and "reconditioned" cars and pickup trucks, imported mainly from Japan. Reconditioned pickups are supplementing and replacing licensed buses and larger lorries to transport people, produce, construction materials, second hand clothes, etc. The pre-dominance of used cars in East Africa produces it own multiplier effect, as import agents, spare parts dealers and auto repair businesses proliferate and prosper.

The use of mass produced goods, such as plastic dishes, aluminum pots, glasses, etc. have displaced many traditional customs and homemade handicrafts, such as using banana leaves as plates or calabashes instead of glasses. The Haya have continuously augmented their consumer purchases since the process began in the 1920s, when farmers first began to participate in a wider market. A well educated Haya, Member of Parliament, who lived in Dar es Salaam, the capital city, lived in a luxury flat, however his family ate their meals on banana leaves in the traditional way — unlike the banana farmers he represented.

Locally Produced Goods

Locally produced goods include furniture (invariably made by local carpenters) *karoboi* (recycled can lanterns) and *ndara* (sandals or thongs made from old rubber tires). This type of production activity is highly creative and innovative; it uses recycled materials, adds value locally and supplies a mass market at affordable prices.

Most small farmers accumulate their inventories of both factory-made and local crafts relatively slowly because they lack ready cash. All regional commodity marketing is highly labour intensive since the buyers are widely dispersed and purchase goods in small lots. So even the sale of imported goods generates a great deal of local employment in the trading sector.

Producer Goods

Two hundred and three of the two hundred and fifty household sample were landholding farmers. Most used very rudimentary farm tools and there is relatively little variation in the number and unit costs of tools between most farms. The standard tools and their replacement costs are listed below: Tibaijuka (1979, p.75) describes and diagrams local cutting knives and spades.

Table 5:
Farm Tools and Livestock (for 203 Sampled Households)

Possession	Percentage	Unit cost (mean $US)
Hoes	97%	6
Panga (machete)	94%	6
Spades	82%	4
Sickle (cutting knives)	78%	5
Cattle manure	54%	15/year
Axes	44%	6
Goats (mean =1.1)	34%	35
Cows (local breed mean =1)	18%	200
(dairy cows mean = 1.9)	6%	400
Wheelbarrows	23%	.6
Chemical insecticide (including cattle dips)	12%	30/year

Average capitalization......................................$176.92

Sixty percent of farmers possess all the basic tools — a hoe, a panga or machete, a spade, a grass cutting knife and a banana cutting knife, plus a few other items on the list above. A small fraction of farmers, six percent, lack even a hoe and a panga; either because they are very old and their farms are in decline or they are very young and their farms are not yet established.

Only relatively well-to-do farmers are cattle owners, with an average of four head of cattle each. The dairy cattle owners are all "rich farmers" (discussed in chapter nine) who own a cattle shed as well as some uncommon items not mentioned on the list.

Most of the products listed above are locally produced (the exceptions are insecticides and one-half the hoes and pangas). Most wheelbarrows are made locally, entirely of wood: they cost a fraction of the price of imported ones. By 1900, cheap hoes and knives imported from Europe almost destroyed a viable thousand year old iron smelting tradition, Kjekshus (1977) and Schmidt (1979). Today there is a revival of local iron working, based on scrap metal use. Almost all of the locally made farm tools are forged from metal scavenged from old trucks, buses, etc.

Very few farmers purchase coffee growing inputs, such as chemical insecticides, in spite of government subsidization. Smallholders distrust chemical insecticides because few receive instruction in their use; almost all farmers relate stories of a neighbour who had lost crops using chemical amendments. Some farmers justifiably fear that if they begin to use insecticides, future supplies will be interrupted, or simply that they will not be able to afford these inputs.

There is a great and growing need for more productive farming technology, but it must be developed to serve local requirements. The most common technological innovations in the rest of the country (oxen ploughs and hybrid maize seeds) are unlikely to make much impact in Kagera region, because ploughs are of little use to local tree farmers. The major economic crops, coffee and bananas, require deep holes for planting seedlings or cuttings; furthermore land topography is uneven and soil depths vary considerably. The landholding system has created a patchwork quilt landscape of small irregularly shaped fenced-off plots. Each of these conditions could prevent plough based agriculture, and in combination they create a formidable barrier.

Real Estate

Haya smallholders spend more money on houses than any other single item: they build in stages as the money for materials and labour becomes available. The traditional circular house with a thatched roof has been replaced by a rectangular one, with roofs made of iron sheets. Only the poorest farmers still use thatch because of leakage, susceptibility to vermin and high maintenance costs.

House building, is no longer a subsistence activity, labour costs are paid in cash or commodities, such as banana beer. Only boys are taught basic skills, and therefore a man building a house, can reduce costs by putting in his own unpaid labour. A woman head of household will have to pay cash for the goods and services listed below.

Table 6:
Expenditure for a Rectangular Dried Mud House 1969 & 1988

(Tanzanian Shs.)	1969	1988
1. Poles and wood	95/-	4000
2. Nails	14/-	1600
3. Planting Poles	1/40	400
4. Iron sheets	360/-	24,000
5. Sheet covers	45/-	5000
6. Sheet nails	21/-	1400
7. Builder's salary	80/-	6000
8. Soil mixer's salary	95/-	3000
9. Carrying vessels & bamboo	10/-	750
10. Six doors	240/-	5000
11. Four windows	40/-	1200
12. The pole collector	19/50	500
13. Food for all builders	20/-	N/A
Total Expenditures	104/TSh	52,350 TSh

Source: Adapted from Rald and Rald (1975, p.102) and personal communication Gration Kabalulu, Ministry of Culture, who was in the process of building a house in his natal village.

In standardized dollar equivalent terms, this archetype house cost $130 in 1969 and $530 in 1988. If the walls of the house are coated with a white adobe-type clay, the construction costs would increase to about $600. If sun dried bricks are used the price would rise to about $700; a cement floor would add another $250, for about 20 bags of cement and labour. To construct this house with walls of locally burnt brick (at 3 Tshs. per brick) would cost about $1300, whereas cement encased mud walls would increase costs to $1600, and walls made of pure cement would increase costs to $2500. The median householder, in the two hundred and fifty household sample, estimated the value of their house at five hundred dollars. The lowest valuation was $100, for an old dilapidated house with no roof sheets, and the highest was $13,000, for a twelve room bungalow with painted cement walls, ceilings and cement floors in every room, and a painted iron-sheet roof. Eight percent of households lived in houses with mud walls and thatched roofs, fifty three percent lived in mud walled houses with iron roof sheets, and thirty nine percent lived in houses with burnt brick or cement walls.

There is a wide gap between the housing enjoyed by the richest fifty-six informant families, living in houses with a median value of $2000, and the poorest thirty-one families living in sub-standard houses which they value at a median of $200. The worst example was a woman farmer and her children living in a house with walls made only from poles, not sealed with mud and therefore exposed to the elements as well as insects, rodents, reptiles and thieves.

Housing expenditure, provides a large amount of regional employment for builders, transporters and traders. Burnt brick production (an important and creative local cottage industry) is an innovative response to the demand for construction material and the high cost of cement blocks. Small producers fire their bricks in a structure that looks like an oven, but is made of sun-dried clay bricks: they burn wood or charcoal inside until the bricks are hard and then take the whole thing apart. Larger brick producers construct their own kilns, so that they can take advantage of an economy of scale. As existing homes deteriorate they need to be renovated or replaced.

There is a large demand for farmland but high prices, ensure that only the richer farmers or those with family connections can buy new plots. Only one-third of the sampled households had been able to purchase a parcel of land — often ten or twenty years ago. Purchased land accounts for only twelve percent of the respondents farmed plots. Sixty-six percent of the rich farmer group purchased land, while only three percent of the poor farmers and nineteen percent of the remainder of households had done so.

Summary

This chapter demonstrates how local traders and local artisans supplied:
1) daily necessities, (despite the description of the Haya as subsistence farmers in much of the literature, in this sample, food was the household's largest expense);
2) consumer durables, such as furniture, radios and bicycles;
3) non-traditional services, such as schools, hospitals and medicines;
4) farm tools and producer goods;
5) construction materials and labour for building and restoring houses.

The real village economy is based on informal sector initiatives: except for sugar, purchased foods are locally produced, as are larger durable goods such as, tables, benches, chairs, and sofas. Most medical care and entertainment is provided by unlicensed practitioners outside the control of the state. About one-half the farm tools are produced locally, as are about three-quarters of construction materials. Production and sale of locally-produced goods, generates cash for local inhabitants of all income levels. In poor countries, such as Tanzania, the small local trader is ubiquitous because of the need to break bulk many times and distribute small lots over a wide area to cash poor consumers.

This chapter demonstrates the importance of the commodity economy and how local production and trading can supply those needs better than large companies or state controlled national initiatives. The local economy is cash based, driven by money rather than barter or pure subsistence. Trade and investment liberalization and structural adjustment should lower the price of roofsheets, sugar and possibly cement even for isolated Haya villagers.

Grassroots innovations in trade and consumer goods signal lifestyle changes and at least the potential for an improved standard of living in the 1990s. Many innovations involve reusing and recycling goods from the outside cultures. Scrap metal becomes agricultural implements, used cans are turned into lanterns, used clothes get worn out, used tires become sandals, newspapers become wrappers, the back of pick-up trucks become buses and on and on. In other cases, villagers have modernized and improved traditional artisanal crafts such as: beer brewing, construction, carpentry, brick making, baking, traditional medicine and midwifery. Both types of innovation demonstrate a dynamic culture capable of problem solving in ways that are *appropriate to local conditions and therefore sustainable.*

In the final chapter, I suggest possible project initiatives which could take advantage of the consumer preferences and purchasing power discussed here. I am convinced that appropriate technical and financial assistance, could improve existing cottage industries that supply food, clothing, household goods and construction materials. Village-based, on-farm enterprises generate significant cash incomes for farmers without adequate land to feed their families. Even if the Haya are likely to remain peripheral participants in the world economy because of factors such as poverty, poor roads and inefficient transport; nevertheless the local economy is dynamic enough to make a difference and improve the overall standard of living.

An enormous population influx and a sizable investment in mineral wealth may also change the status quo and compel the government to build up the transport and communications infrastructure. During the April 1994 to December 1996 period, Tanzania hosted almost a million Rwandan refugees, who came in two waves, Tutsis during the genocide and Hutus fleeing the RPF invasion later in 1994. The camps have now been largely abandoned but under UN sponsorship a significant amount of work was done to improve the infrastructure in the area of the camps and the routes leading up to the camps. However much of this work was of a temporary nature and may quickly deteriorate and as noted in chapter six, many trees were cut down and will take some time to replace. Also a natural disaster, the El Nino rains caused massive flooding in the region The other possible change factor is the proposed foreign investment in large-scale commercial gold mining. This could mean potentially dramatic improvements in roads, power, telephone service and other infrastructure advancement, since gold mines would create local employment, generate taxes and necessitate new capital installations of all sorts.

APPENDIX: PURCHASES OF DAILY NEEDS (28 HOUSEHOLD SUB-SAMPLE: JANUARY 1 TO JUNE 1 1989)

Numbers represent Tanzanian shillings at a time when $1 US = 50 Tsh.

BANANAS........................ 590, 1280, 735, 0, 0, 20, 110, 0, 220, 545, 0, 0, 365, 0, 200, 580, 470, 0, 930, 0, 470, 0, 990, 200, 200, 0, 140, 0.

BANANA BEER & LIQUOR.......... 0, 1100, 800, 800, 190, 200, 0, 0, 0, 85, 420, 0, 150, 220, 170, 430, 750, 100, 355, 1350, 550, 290, 755, 40, 0, 0, 5310, 90.

BATTERIES...................... 0, 170, 280, 0, 0, 0, 0, 400, 0, 0, 150, 120, 0, 0, 0, 0, 0, 0, 120, 0, 0, 100, 0, 450, 200, 0, 0, 0.

BEANS......................... 150, 0, 0, 0, 45, 0, 0, 450, 0, 20, 0, 0, 0, 50, 0, 0, 0, 0, 0, 0, 0, 0, 0, 135, 35, 50, 20, 0.

BLANKETS & BED SHEETS & TOWELS.630, 500, 0, 500, 400, 0, 400, 600, 0, 0, 0, 300, 0, 0, 0, 0, 0, 0, 160, 300, 0, 0, 250, 0, 0, 0, 0, 300.

BREAD......................... 340, 0, 590, 380, 0, 120, 0, 250, 480, 100, 100, 100, 0, 220, 320, 0, 0, 0, 100, 0, 0, 300, 0, 120, 120, 0, 0, 100.

CAKES & SNACKS................ 168, 140, 280, 80, 125, 0, 30, 530, 180, 70, 75, 100, 0, 0, 70, 20, 5, 0, 50, 0, 35, 0, 0, 110, 150, 0, 60, 120.

CIGARETTES & TOBACCO.......... 0, 0, 0, 50, 0, 0, 0, 0, 0, 0, 0, 0, 40, 0, 290, 400, 15, 0, 30, 220, 100, 0, 310, 210, 50, 60, 340.

CLOTHES....................... 3000, 1530, 2200, 1350, 1580, 2100, 1580, 13,300, 1200, 0, 0, 2800, 770, 142, 600, 800, 2850, 230, 2490, 300, 2700, 1100, 40, 2830, 1530, 0, 1210, 450.

COOKING OIL 60, 410, 250, 400, 260, 0, 0, 280, 125, 15,

350, 450, 360, 190, 420, 220, 220, 40, 60, 110, 120, 390, 0, 540, 350, 0, 320, 300.

COSMETICS & JEWELLERY........... 0, 60, 163, 240, 0, 0, 260, 235, 0, 0, 0, 180, 0, 0, 0, 0, 0, 0, 0, 0, 0, 0, 0, 60, 60, 120, 0, 180.

DAGAA (dried sardines)......................... 220, 220, 80, 50, 145, 60, 0, 235, 0, 60, 92, 240, 160, 0, 300, 90, 200, 70, 240, 130, 200, 50, 205, 240, 140, 80, 175, 200.

FIREWOOD & CHARCOAL........... 20,140, 540, 150, 60, 60, 180, 240, 800, 180, 0, 0, 50, 0, 50, 150, 60, 0, 0, 0, 60, 35, 0, 85, 85, 0, 510, 25.

FLOUR......................... 310, 940, 1420, 730, 125, 170, 50, 1270, 300, 560, 365, 200, 120, 300, 220, 320, 150, 360, 1130, 430, 100, 220, 810, 270, 200, 180, 160, 210.

FRESH FISH.................... 1090, 1240, 365, 305, 2115, 1235, 120, 1145, 210, 410, 1755, 905, 700, 100, 910, 1090, 940, 240, 1095, 1150, 700, 1075, 1095, 1665, 1129, 615, 1370, 745.

FRUIT......................... 0, 0, 30, 0, 0, 0, 0, 300, 0, 10, 70, 0, 0, 0, 0, 0, 0, 0, 5, 0, 0, 0, 0, 50, 50, 0, 0, 0.

KAROBOI ....0, 0, 200, 0, 400, 25, 0, 0, 400, 40, 0, 0, 170, 110, 0, 0, 0, 0, 100, 500, 0, 0, 0, 60, 60, 0, 0, 30.

KEROSENE...................... 470, 480, 280, 260, 770, 285, 150, 620, 125, 30, 390, 195, 165, 210, 270, 440, 120, 135, 160, 440, 220, 390, 180, 430, 320, 180, 515, 800.

KITCHEN WARE................. 442, 180, 2100, 95, 650, 700, 70, 500, 0, 200, 0, 70, 50, 220, 0, 600, 300, 0, 230, 300, 300, 490, 250, 365, 150, 0, 330, 400.

MATCHES...................... 40, 55, 10, 10, 210, 40, 10, 140, 20, 10, 30, 40,

20, 30, 85, 60, 0, 10, 30, 20, 0, 45, 20, 30, 20, 30, 10, 40.

MEAT.......................... 440, 1090, 440, 1435, 640, 580, 170, 1780, 830, 405, 200, 180, 370, 580, 270, 310, 550, 120, 300, 455, 370, 750, 0, 1120, 800, 0, 540, 580.

MEDICINE....................... 30, 90, 30, 260, 30, 40, 10, 130, 120, 18, 15, 45, 50, 50, 10, 130, 330, 410, 0, 20, 330, 30, 40, 350, 70, 22, 90, 90.

MILK.......................... 125, 120, 190, 0, 200, 0, 0, 260, 80, 0, 40, 0, 30, 0, 0, 68, 120, 0, 80, 40, 40, 100, 0, 90, 90, 0, 30, 50.

NEEDLES & THREAD............... 40, 20, 10, 0, 0, 25, 30, 0, 0, 8, 0, 0, 20, 0, 9, 76, 15, 0, 18, 0, 15, 15, 0, 125, 75, 5, 22, 0.

POTS & PANS................... 0, 300, 1700, 0, 300, 0, 0, 0, 0, 350, 0, 350, 0, 800, 0, 0, 0, 0, 0, 0, 0, 0, 0, 0, 0, 0, 350, 0.

RICE.......................... 280, 610, 1000, 900, 500, 770, 100, 1100, 0, 650, 560, 430, 160, 130, 50, 220, 0, 0, 140, 0, 0, 710, 0, 320, 440, 0, 45, 500.

SALT.......................... 290, 220, 0, 0, 45, 50, 0, 150, 0, 20, 30, 60, 80, 190, 190, 70, 45, 80, 150, 195, 145, 220, 50, 140, 140, 80, 130, 50.

SHOES & SHOE POLISH........... 0, 1200, 50, 0, 0, 0, 400, 2000, 0, 0, 0, 0, 0, 0, 1600, 500, 0, 0, 500, 0, 0, 0, 0, 200, 0, 0, 0, 0.

SOAP & CLEANERS............... 110, 250, 470, 320, 860, 416, 130, 1160, 210, 155, 220, 260, 255, 250, 510, 200, 350, 250, 100, 310, 230, 300, 330, 585, 262, 100, 495, 260.

SOFT DRINKS................... 190, 330, 240, 0, 0, 1335, 0, 0, 0, 0, 0, 30, 0, 0, 0, 0, 30, 30, 0, 0, 0, 100, 100, 0, 100, 0.

SPICES........................ 0, 70, 0, 100, 20, 0, 0, 50, 0, 30, 0, 0, 0, 0, 0, 0, 0, 0, 20, 0, 0, 0, 0, 0, 0, 0, 40.

STATIONERY.................... 76, 270, 160, 209, 130, 15, 10, 175, 0, 25, 0, 180, 20, 60, 15, 20, 15, 0, 80, 15, 15, 105, 0, 85, 80, 30, 78, 55.

SUGAR.......................... 2070, 2780, 1250, 200, 1320, 1100, 400, 5000, 500, 610, 900, 1345, 1168, 800, 1000, 1210, 790, 400, 670, 300, 790, 190, 900, 1770, 1200, 300, 680, 1320.

TEA & COFFEE.................. 214, 160, 170, 0, 55, 50, 30, 105, 0, 25, 0, 90, 90, 0, 0, 30, 30, 50, 30, 70, 30, 90, 130, 205, 70, 0, 48, 70.

TUBERS......................... 0, 0, 165, 160, 0, 0, 0, 0, 160, 0, 0, 0, 0, 200, 60, 180, 80, 0, 0, 0, 80, 0, 300, 0, 0, 0, 0, 0.

VEGETABLES.................... 30, 120, 555, 170, 180, 20, 50, 1050, 200, 80, 120, 0, 0, 0, 0, 0, 40, 80, 30, 0, 10, 100, 10, 190, 150, 10, 20, 25.

WASH BASIN & BUCKET........... 0, 0, 0, 200, 400, 0, 0, 0, 300, 140, 350, 0, 0, 0, 0, 0, 250, 0, 0, 0, 250, 0, 0, 0, 0, 0, 0, 0.

OCCASIONAL PURCHASES (28 HOUSEHOLD SUB-SAMPLE JANUARY 1 TO JUNE 1 1989)

AXE @ 600:

BICYCLES @ 10,000, 10,000:

BICYCLE PARTS @ 350, 50, 350, 1000:

BRICKS @ 17,500, 2500:

BUS FARE @ 400:

CARRYALL BAGS @ 2000, 2800, 3800:

CEMENT	@ 6000:
CHARCOAL COOKER	@ 400:
CHICKEN	@ 300, 300:
COFFEE DRYING MAT	@ 300, 230:
CUPBOARD	@ 2000:
CURTAIN	@ 400:
CUSHION	@ 250:
DENTIST FEE	@ 1200:
EYE GLASSES	@ 500:
FISH HOOK	@ 15, 15:
GARDEN SPADE	@ 500, 200:
GERRY CAN	@ 140:
GOATS	@ 3500:
GUNNYSACK	@ 40, 500, 2000:
HOE	@ 640, 450:
IRON ROOFSHEET	@ 1500:
KEROSENE STOVE	@ 1500:
LATRINE	@ 700:

LUMBER	@ 2000, 2000, 700, 1000:
MANURE	@ 600, 60, 100:
MATTRESS	@ 3500, 4000:
MIRROR	@ 120, 200:
PADLOCK	@ 700, 160:
PANGA	@ 400, 480, 400, 350:
PASTA	@ 200:
PHOTOGRAPH (of family)	@ 400, 200:
RADIO	@ 2000, 3000:
RADIO CASSETTE	@ 4,000:
ROPE	@ 600, 350, 500, 300:
SEEDS & SEEDLINGS	@ 200, 200:
SICKLE	@ 170, 600:
STATUE OF VIRGIN MARY	@ 400:
SUITCASE	@ 3800:
TABLE	@ 300:
THERMOS FLASK	@ 700:
TOOLS	@ 300, 15,000, 900:

TRAIN FARES	@ 210:
VIDEO SHOW	@ 200, 200:
WATCH REPAIR	@ 200, 200:
WOODEN WHEELBARROW	@ 500, 500:

12 Conclusion

The Haya people have been the subject of research discourses since the expeditions of Burton, Speke and Stanley in the late nineteenth century and the ethnography of father Césard in the early years of this century. Cory and Hartnoll codified Haya customs in their voluminous work in the 1940s republished in 1971. Most recently Anna Tibaijuka and Magdelena Ngaiza have carried out detailed research. My empirical findings are similar to those of Tibaijuka and Ngaiza but I investigated an even more extensive network of commodity relations at the village level. Village production satisfied almost all the local demand for goods and services at the beginning of the 1990s but with trade liberalization and increasing foreign investment in the late nineties the situation may be changing: this is a topic that needs to be researched.

If my work and previous and forthcoming research are to serve the Haya people, then there are three main challenges that need to be met: empowering women farmers, reducing alcohol abuse, and providing poorer villagers with start-up capital.

Land inheritance based on customary law and patriarchal customs, such as bride price, force women to put in extra hours and make male labour the only available underutilised resource. Woman heads of household support their families on smaller plots of land with fewer resources and therefore males could improve productivity and women (given more land and capital), could live more comfortably.

Can banana brewing and distilling lead to development? Despite the Government's sporadic attempts to stop the *konyagi* trade it is a growth industry that I estimate produces 10 million US dollars worth of revenue in the region. This puts brewing on a par with coffee in terms of generating revenue, the challenge will be to control some of the negative side affects of the alcohol trade which result from alcoholism and related social problems. Stakeholders who claim to be "alleviating poverty" must try controlling social problems associated with hard liquor, making it more sanitary and generating indirect revenue, since it cannot be stopped or directly controlled.

This study demonstrated that becoming a rich farmer was not primarily a question of access to modern education, adopting a "scientific" world view or a stronger work ethic. Of course these attributes can help, but the foremost factor determining material success is access to strategic capital. Enterprise and managerial skill are prerequisite to material success but only after adequate capital is accumulated and put to productive use.

Coffee is Kagera Region's only notable export. In the early 1990s the KCU, Kagera Cooperative Union, made annual payments of about 10 million US dollars to coffee farmers. Assuming that my sample is fairly typical of the region then about 50 million US dollars per year is spent on trade in locally produced goods. More work needs to be done to sketch a composite picture of the informal economy, but the type of grassroots innovations described in the book could provide a model for projects to promote greater equality and prosperity for all.

Despite the fact that most researchers describe Kagera's economy as subsistence-based, there was ample evidence of a cash nexus as village traders and middlemen supplied goods and services in exchange for payment in at least a dozen enterprises. To refresh the reader's memory I will list the key innovations in the ways the Haya clothe, feed, entertain themselves and earn a living. These are the most successful sectors of the local economy where there is room for significant expansion.

1. Produce
(includes fresh fish, dried smoked minnows, maize/cassava flour, garden vegetables, plantains, milk, meat, eggs, root crops and fruit)
2. Woodworking & Furniture Making
(includes cut lumber, seats, benches, tables, beds, sofas, armchairs, wooden wheelbarrows, mortar and pestles)
3. Fuel Related
(includes firewood and charcoal produced by smallholder agro-foresters, clay ovens)

4. Construction
(includes fired clay bricks, cutting and installing poles, white and red clay plastering, sun-dried and kiln-dried bricks, thatching, window making, masonry and cement work, roof and ceiling installation, gutters, cisterns and water collection systems)
5. Catering and Food Processing
(includes roasted peanuts, chewing coffee, cakes, bread baked in heated sand or brick ovens, samosas, grinding mills for grain and beans: mills can eliminate the time women spend pounding grains to make flour)
6. Tailoring
(includes dresses, school uniforms, embroidery, and design and customized production of clothes)
7. Handicrafts
(includes tire sandals, batik, tie-dyes, barkcloth, carving, card and envelope making, basket and mat weaving)
8. Metal Working
(includes farm tools forged from scrap iron, *karoboi* small lamps made out of soldered cans)
9. Transportation
(in-village porterage using bicycles and wheel barrows to replace head loads)
10. Traditional Medicine and Herbalism
(there is a pressing need to document, test and standardize a veritable treasure trove of existing botanical knowledge and herbal lore and to protect threatened species and perhaps discover plants with export potential in the pharmaceutical sector).[1]
11. Brewing and Distilling
(production of special banana varieties, new ripening methods, new stills)
12. Dairy Production
(using non-traditional methods of stall feeding and purebred cattle).

All the above goods are being successfully manufactured in the village environment: wouldn't these activities foster even more growth if financial assistance, training and encouragement were provided to disadvantaged residents? Shouldn't richer farmers be enlisted as trainers and animators, each teaching his or her own specialty? The pay-off for trainers would be social prestige and a share in an expanding local economy.

Should projects which seek to communalize larger capital goods, such as, buses, trucks, lorries, tractors, etc., be avoided in the present conjuncture? In Tanzania, the larger cooperative ventures set up as national programmes, have tended to stagnate, whereas, private initiatives (mainly in the informal sector) have done much better. Is it possible to ensure that projects which require expensive equipment remain democratic and incorrupt, or are these best left to a later stage of development?

Some development projects succeed because they tie into Haya culture; the best example is the dairy cattle initiative. The improved-stock dairy cattle project demonstrates that the Haya will eagerly turn to on-farm innovations that tie in with their way of life and provide prestige (accorded to cattle owners) profit (based on cash sales of milk) farm capital (manure) subsistence (milk) and environmental relief (for the overgrazed pasture land). The main disadvantage of the dairy cattle project is the initial high cost of cows and the related on-going costs such as, stalls, feed, and veterinary care. Can other assistance projects, targeting the poor and middle status farmers, succeed equally well? Can they integrate into the local pattern of production, trade, and consumption? What is the prognosis for regional economic development based on the grassroots innovation model explored in previous chapters?

Village cottage industries already produce most basic commodities. Can activities which add value to local raw materials be encouraged with training, technical assistance and loan capital? Can women and poor young men benefit from appropriate assistance that helps them set up viable enterprises that do not require much start up capital? I hope the answer to these questions is positive because the fate of not only the Haya but many other Africans may rest on the success of grassroots innovations. Little else has worked so far!

Can national organizations, and the new wave of corporate investors cooperate with localized farmer-initiated innovations in the sectors listed above? Can such projects raise real incomes by providing not only cash but consumer goods on a regular basis and at low cost? Ultimately if the Tanzanian state is to overcome its continuing "fiscal crisis" and succeed in its proposed tax reforms, it must encourage the peasants to increase their involvement in the official economy and provide material incentives for them to do so. But the agents of the aid donors and the government may need a push from the popular bloc of civil society that will provide the social force, the political will and the economic means so that the system does become more accountable and an "enabling environment" is created and the nation can prosper and resolve its enormous economic problems.

Civil society in the village context refers to informal gatherings rather than registered associations. This study indicates that the best way to use civil society as a tool for democratization and development is through incentive programs that literally deliver the goods. As many commentators have already pointed out only improved economic performance will sustain and reinforce both political and economic reforms. The village communications networks are very effective when it comes to selecting useful innovations and screening out and ignoring useless ones. This study has demonstrated many examples that contradict the *assumption* that peasants are inherently too conservative to innovate: they will readily do so when it is in their interest.

Note

1 For example a rainforest tree known as "neem" is quite common in Africa; it is being tested and developed as an inhibitor of the HIV virus, by the US government's National Cancer Institute. This plant is known as arobaini (forty in kiswahili) because it is used as a traditional medicine in treating forty ailments, Meanwhile, researchers at New Delhi's National Institute of Immunology have begun investigating the anti-HIV properties of at least eight plants. (WORLD AIDS MAGAZINE, 1993)

Bibliography

Adelman, H. and Sorenson, J. (1994) *African Refugees: Development Aid and Repatriation,* Westview Press, Boulder.

Aké, C. (1991) "Rethinking African Democracy", *Journal of Democracy*, vol. 2/ 1.

Allan, W. (1965) *The African Husbandman* Oliver & Boyd, Edinburgh.

Almond, G. and Verba, S., (1963) *The Civic Culture,* Princon University Press, Princeton.

Amin, S, (1987) "Preface: The State and the Question of Development" in P. A. Nyang'o (ed.) *Popular Struggeles for Democracy in Africa,* Zed Press, London.

Arhem, K. (1986) "Pastoralism Under Pressure: The Ngorongoro Masai" in Boesen et al., *op cit.*

Aronowitz S. and Giroux H. (1993) *Education Stil Under Siege,* OISE Press, Toronto.

Austen, R.A. (1968) *Northwest Tanzania Under German and British Rule:Colonial Policy and Tribal Politics, 1889-1939*, Yale University Press, New Haven.

Austen, R.A. (1965) *Native Policy and African Politics: Indirect Rule in Northwest Tanzania, 1889-1939*, Ph.D. Dissertation, Harvard University.

Awiti, JR. (1972) "Class Struggle in Rural Society of Tanzania", *Maji Maji,* No.7.

Bader, Zinnat K. (1975) *Women, Private Property and Production in Bukoba District,* University of Dar es Salaam, M.A. dissertation.

______ (1982) *Women, Landownership and Ideology in Bukoba District,* mimeo, Cambridge, Mass.

Baines, D.L. (undated) "Notes on German Administration" in *Cory Collection,* University Library, Dar Es Salaam.

Bakenges, S.K.S (1974) *An Historical Survey of the Coffee Industry in Bukoba District,* M.A. Dissertation, Universtiy of Dar es Salaam.

Barkan, J. D. (1994 "Divergence and Convergence in Kenya and Tanzania: Pressures for Reform" in J. D. Barkan (ed.) (1994) *Beyond Capitalism versus Socialism in Kenya and Tanzania,* East African Educational Publishers, Nairobi.

Barker, J. S. (1994) "Local Political Space" presented at the annual meetings of the African Studues Association, Toronto.

Bartlett P.F. (1980) "Adaptive Strategies in Peasant Agricultural Production" *Annual Review of Anthropology* 9: pp. 554-73.

Bauer P. T. (1974) *Nigerian Development Experience: Aspects and Implications,* Ife-Ife Nigeria: University of Ife Press.

Bay, Edna, ed. (1982) *Women and Work in Africa*, Westview Press, Boulder.

Bayart, J-F. *The State in Africa: The Politics of the Belly,* Longmans, London.

Beattie J. & Middleton J. (eds.), *Spirit Mediumship and Society in Africa,* London.

Beckman B., (1988) "The Post-colonial State: Crisis and Reconstuction", IDS Bulletin,vol. 19/ 4.

Berg, E. (1981) *Accelerated Development in Sub-Saharan Africa: An Agenda for Action,* World Bank Report.
Berg, R. and Whitaker, J. (1986) *Strategies for African Development*, University of California, Berkeley.
Berger, R. (1963) "Oral Traditions in Karagwe," *mimeo, U.iversity of Dar es Salaam.*
Bermeo, N., (1992) "Rethinking Regime Change" in *Comparative Politics* vol. 24/ 3.
Bernstein, Henry (1976) "Notes on Capital and Peasantry" *Economic Research Bureau Occasional Paper,* Dar es Salaam.
______ (1977) "Notes on Capital and Peasantry" in *The Review of African Political Economy* No. 10.
______ (1979) "African Peasantries: A Theoretical Framework" *Journal of Peasant Studies,* Vol. 6, No.4.
______ (1981) "Notes on State and Peasantry: The Tanzanian Case" *The Review of African Political Economy* No. 21.
Berry, Sara S. (1985) *Fathers Work for their Sons*, Westview Press, Boulder.
_____ (1986) "Economic Change in Contemporary Africa" in *Africa,* edited by Phyllis Martin and P. O'Meara, Bloomington.
Boesen J. (1973) "Peasants and Coffee Export: A Coffee Exporting Region in Tanzania" in *Dualism and Rural Development in East Africa,* Institute for Development Research, Copenhagen.
Boesen J., Madsen,B.S., Moody I., (1977) *Ujamaa: Socialism from Above,* The Scandinavian Institute of African Studies, Uppsala.
Boesen J., Havnevik,K.J., Koponen,J,. Odgaard, R., editors, (1986) *Tanzania: Crisis and Struggle for Survival,* SIAS Uppsala.
Bohannan, P.(1959) "The Impact of Money on an African SubsistenceEconomy", *The Journal of Economic History,* 19 (4).
_____ (1964) "Land Use, Land Tenure and Land Reform" in *Economic Transition in Africa,* eds. Melville Herskowitz andMitchell Harwitz, Northwestern U.P., Chicago.
Bolton, Dianne (1985) *Nationalization: a Road to Socialism,* Zed Press, London.
Boserup, Esther (1970) *Woman's Role in Economic Development,* Allen and Unwin, London.
Botte, R. (1974) "Processus de Formation d'une classe sociale dans une societé Africains précapitaliste" *Cahiers d'études Africaines* Vol.14, No. 4: pp.605-626.
Boulaga, E. (1993) *Les Cnferences Nationales en Afrique Noire,* Editions Kathala, Paris.
Bourdieu, P. (1977) *Outline of a Theory of Practice,* Cambridge University Press.
Bowles B.D. (1979) "The Political Economy of Colonial Tanganyika 1939-1968" in M.H.Y. Kaniki ed. *Tanzania Under Colonial Rule,* London, Longmans.
Brain, J.L. (1973) "The Tutsi and the Ha:" A Study of Integration" *Journal of Asian and African Studies,* Vol.8, No. 1-2; pp.39-49.
Bratton, M. (1989) "Beyond the State: Civil Society and Associational Life in Africa" in *World Politics* vol. 41/ 3.

______ (1994) "Micro-Democracy? The Merger of Farmer Unions in Zimbabwe" in *African Studies Review* vol 37/ 1.

____, and van de Walle, N. (1994) "Neo-Patrimonial Regimes and Political Transitions in Africa" *World Politics* vol. 46/ 4.

Brett, E.A. (1973) *Colonialism and Underdevelopment in East Africa*, London.

British Sessional Papers, (1922), House of Commons, Vol. XVI,
(Report on Tanganyika Territory, (1931/32), Volume VII,
Armitage-Smith, S.,(1921) *Report on a Financial Mission to Tanganyika* Vol. XXIV .

Bryceson D.F. & Mbilinyi M.(1978) *The Changing Role of Tanzanian Women in Production: From Peasants to Proletarians*, DSM:UDSM Bureau of Resource Assessment and Land Use Planning BRALUP Service Paper 78/5.

Bryceson Deborah F. (1979)"Primitive Accumulation and Imperialism in Relation to the Reproduction of Third World Peasantries"unpublished paper presented at the Southern African Universities Social Sciences Conference, Dar es Salaam Tanzania, June.

____ (1980) "Changes in Peasant Food Production and Food Supply in Relation to the Historical Development of Commodity Production in Pre-Colonial and Colonial Tanganyika" *Journal of Peasant Studies*, Vol.7,No.3.

_____ (1990) *Food Insecurity and the Social division of Labor in Tanzania, 1919-85*, St. Martins Press, New York.

Bugengo, J., Mutangira, J.P.B. and Rwelegera, J.B.K. (1976)"The Nyarubanja System and Ujamaa Village Development in West Lake Region" *Economic Research Bureau Paper,* 76, 1, Dar es Salaam.

Bugra, Janet (1975) "Women Entrepreneurs of Early Nairobi" *Canadian Journal of African Studies* vol. 9 no. 2.

Bunker, Stephen *Peasants Against the State: The Politics of Market Control in Bugisu Buganda, 1900-1983,* Urbana: University of Illinois Press.

Callaghy, T. (1987) "The State as a Lame Leviathon: The Patrimonial Administrative State in Africa" in Z. Ergas (ed.) *The African State in Transition,* Macmillan, London.

______ (1991) "Africa and the World Economy: Caught Between a Rock and a Hard Place" in J. Harbeson, and D. Rothchild (eds.) *Africa in the World Economy,* Westview, Boulder.

____ (1994) "Civil Society, Democracy and Economic Change in Africa: A Dissenting Opinion about Resurgent Societies" in J. Harbeson, D. Rothchild and N. Chazan (eds.) *Civil Society and the State in Africa,* Lynne Rienner, Boulder.

Cameron, Sir D. (1926) "Agriculture and Labour", *East Africa Pamphlet,* Colonial Office Library No. 10942.

Caplan, Pat (1981) "Development Policies in Tanzania: Some Implications for Women" in Nici Nelson *African Women in the Development Process,* London; Frank Cass.

Carrothers, Thomas (2000) "Civil Society" *Foreign Policy,* Winter 199-2000.

Cecelski E. (1984) *The Rural Energy Crisis, Women's Work and Family Welfare: Perspectives and Approaches to Action,* ILO, Geneva.

Cernea, Michael M (1985) *Putting People First: Sociological Variables in Rural Development,* Washington: Published for the World Bank by Oxford University Press.

Césard, E. (1937) "Les Muhaya", *Anthropos,* Vol.XXXII.

Chanock M. (1985) *Law, Custom and Social Order: The Colonial Experience in Malawi and Zambia,* Cambridge University Press.

Chatterjee, P. (1990) "A Response to Taylor's 'Modes of Civil Society'," in *Public Culture* vol. 3/ 1.

Chave-Fallers, M. (1960) *The Eastern Lacustrine Bantu,* London.

Chazan, N. (1992) "Africa's Democratic Challenge" *World Policy Journal* vol. 19/ 2.

Chege, M. (1994) "The Return of Multiparty Politics," in Barkan, *op cit.*

Chilver,E.M. (1959) "Feudalism in the Interlacustrine Kingdoms"in Richards, R.I. *East African Chiefs,* London.

Chittick, H.N. and Rothberg, R.I. (eds.)(1975) *East Africa and the Orient*, Africana, New York.

Cliffe, L. and Saul, J. (eds.) (1972) *Socialism in Tanzania,* Nairobi.

Cliffe, Lionel (1987) "The Debate on African Peasantries" *Development and Change* 18 (2).

Cohen, J. L, and Arato, A. (1992) *Civil Socity and Political Theory,* MIT Press, Cambridge.

Coleman, J.S. (1968) "Modernizaion, Political Aspects" in *Internation Encyclopedia of the Social Sciences* vol, 10, Macmillan, New York.

Colonial Office Library *Tanganyika, Original Correspondence, and Other Named Documents.*

Colonial Office (1950) "The Lake Province Council", *Journal of African Administration* vol. 2, No.1.

____ (1951) E.F. Twinning (Governor) *Development of African Local Government in Tanganyika,* HMSO, London.

Cooksey,B,. Fowler, A., and. Kwaya, C., (1986) *Incentive Goods for Development: A Survey of Demand Priorities Among Villagersin Tanga and Mwanza Regions and Moshi and Hai Districts,* Nairobi, Consultants for Management Development Programs.

Cory, H. (1942)"Tribal Structure in Uhaya" *Tanganyika Notes and Records* No. 14.

______ (1948) *The Haya Tribe and the Incidence of Venereal Disease* Cory Collection No. 104, University Library, Dar es Salaam.

______ (1954) *The Indigenous Political System of the Sukuma,* Dar es Salaam.

_____ (1955) *A Report on the Nyarubanja System in Bukoba District,* Cory Collection No. 72, University Library, Dar es Salaam.

____ (undated) *The Bahinda in Bumbwiga,* Cory collection No. 40,University Library, University of Dar es Salaam.

_______ (1956) "Bahaya and the African Explorer" *Tanganyika Notes and Records*

No.43.

Cory H. & Hartnoll M. (1971) *The Customary Law of the Haya Tribe* Frank Cass, London.

Coulson, A. (1977) "Agricultural Policies in Mainland Tanzania", *Review of African Political Economy,* No.10.

Coulson, A. (1982) *Tanzania: A Political Economy*. Oxford, Oxford University Press.

Cowen, M. (1981) "Commodity Production in Kenya's Central Province" in ed. Heyer,G. et al, *Rural Development in Tropical Africa,* London.

Crook, R.E. (1990) "State Society and Political Institutions in Cote D'Ivoire and Ghana" *Institute of Development Studies Bulletin* vol.21/ 4.

Crush, Jonathan (1988) "Brewing for Skroof" Studies Association Annual Meetings, Chicago.

Curry R.L. (1988) "Alcohol and African Underdevelopment" a paper presented to the African Studies Association Annual Meetings, Chicago.

Curtis Kenneth R. (1989) "Cooperation and Co-optation: The Struggle for Market Control in the Bukoba District of Colonial Tanganyika" *International Journal of African Studies East Africa,* Vol. 5.

Dahl, R. A. (1971) *Polyarchy: Participation and Opposition,* Yale University Press, New Haven.

Diamond, L., Linz, J. and Lipset, S. M. (1988) *Democracy in Developing Countries: Africa,* Lynne Rienner, Boulder.

Di Palma, G. (1990) *To Craft Democracies, An Essay on Democratic Transitions* University of California Press, Berkeley.

Doro, M.E. (1996) "Democratization in Africa: a Preliminary Assessment" in *African Voices: A Newsletter on Democracy and Governance in Africa,* VOL. 5/ 2 USAID.

Eicher, Carl K. and J.M. Staatz (1984) *Agricultural Development in the Third World* Baltimore:, Johns Hopkins University Press .

Elam, Y. (1974) "The Relationships Between Hinda and Iru in Ankole" *African Studies* Vol.33,No. 3: pp.159-172.

_____ (1973) *The Social and Sexual Roles of Hima Women,* Manchester.

Ellis, Frank (1982) "Agricultural Price Policy in Tanzania," *World Development,* 10(4).

Etienne M., & Leacock E.B. (1980) *Women and Colonization* New York.

Evans, P., Rueschemeyer, D. and Stephens, E.H., (1985) *State Versus Markets in the World System,* Sage, Beverley Hills.

Eyoh, D. (1996) "From Economic Crisis to Political Liberalization: Pitfalls of the New Political Sociology for Africa" in *African Studies Review* vol. 39/ 3.

Fallers, L. (1956) *Bantu Bureaucracy,* London.

____ (1961)"Are African Cultivators to be Called Peasants?" *Current Anthropology,* Vol.,12, No.2.

F.A.O. (1961) *The World Coffee Economy,* Commodity Bulletin Series No.33.

_____ (various years) *Production Yearbooks.*

Fairclough J.L. (undated) *Report on the Nyarubanja of Land Tenures in the Bukoba*

District TNA file number 1418.

Fatton, R. (1992) *Predatory Rule: State and Civil Society in Africa,* Lynne Rienner, Boulder.

______ (1995) "Africa in the Age of Democratization; The Civic Limitations of Civil Society" in *African Studies Review* vol. 38/ 2.

Flint, J. (1963) "The wider Background to Partition and Colonial Occupation" in Oliver and Matthew *History of East Africa,* Oxford.

Ford, J. and Hall, R. de Z. (1947) "The History of Karagwe (Bukoba District)" Tanganyika Notes and Records, No.24.

Ford, V.C.R. (1955) *The Trade of Lake Victoria,* Kampala.

Fortmann, Louise (1982) "Women's Work in a Communal Setting: The Tanzanian Policy of Ujamaa," in E. Bay, *Women and Work in Africa.*

Freeman-Grenville, G.S.P. (1963) "The German Sphere 1884-1898" in Oliver and Matthew *History of East Africa,* Oxford.

Friedrich K.H. (1968) "Coffee-Banana Holdings at Bukoba" in Hans Ruthenberg ed. *Smallholder Farming and Smallholder Development in Tanzania: Ten Case Studies,* Weltforum Verlag: Munchen.

Fuggles-Couchman, B.R. (1964) *Agricultural Change in Tanganyika 1945-1960,* Stanford.

Fuigitt, Diana (1985) "The Post-Arusha Development of Economic Organization in Tanzania's Agricultural Sector" presented at the African Studies Association annual meetings, New Orleans.

Galbraith, G.S. (1972) *Mackinnon and East Africa: 1878-1895*, Cambridge U.P.

Geertz, C. (19900 "Blurred Genres: The Reconfiguration of Social Thought" in *American Scholar* vol. 49.

Gerschenberg, I. (1971) "Customary Land Tenure as a Constraint on Agricultural Development: A Re-Evaluation" *East African Journal of Rural Development,* Vol. 4, No. 1.

Gifford, P. and Lewis, W.R. (1967) *Britain and Germany in Africa,* New Haven.

Gladwin, Christina H., K. Staudt, and D. McMillan (1986) "Providing Africa's Women Farmers Access: One Solution to the Crisis" *Journal of African Studies,* 13 (4).

Glazier J. (1985) *Land and the Uses of Tradition among the Mbeere of Kenya,* The University Press of America, Lanham.

Gold, T. B. (1986) *State and Society in the Taiwan Miracle,* M.E. Sharpe, New York.

Goody, J. (1969) "Economy and Feudalism in Africa", *The Economic History Review*, Vol. XXII, No.3.

_____ (1971) *Technology Tradition and the State in Africa*, London.

Graham, James (1968) *Changing Patterns of Wages in Tanzania,* Ph.D. Dissertation, Northwestern University, Chicago.

Gray, Sir J. (1957) "Trading Expeditions from the Coast to Lakes Tanganyika and Victoria Before 1857 " *Tanganyika Notes and Records* T.N.R., No. 49.

Griffith, A.W.. (1936) "Primitive Native Eduction in Bukoba District" *Tanganyika Notes and Records,* No.1.

Gulliver, P.H. (1958) *Land Tenure and Social Change Among the Nyakusa,* Kampala.
Guyer, Jane I. (1980 "Female Farming and the Evolution of Food Production Patterns Amongst the Beti of South-Central Cameroon" *Africa* 50 (4).
(1986) "Women's Role in Development", in R. Berg and J.S. Whitaker *Strategies for African Development*, University of California Press: Berkeley, Los Angeles.
Gwynne, M.D. (1975) "The Origin and Spread of Some Domestic Food Plants of Eastern Africa" in Rothberg and Chittick *op cit*.
Haarer H.E. (1956) *Modern Coffee Production,* London.
Hall, R.de Z. (1929) "Ethnological and Anthropological Notes" T.N.A. (Bukoba District Book.)
Hancock, G. (1989 & 1991) *Lords of Poverty,* Mandarin Press, London.
Harlow, V. and Chilver, E.M. (1965) *History of East Africa,* Oxford.
Harrison, M. (1977) "The Peasant Mode of Production in the Work of A.V. Chayanov" *The Journal of Peasant Studies,* Vol.4, No. 4.
Harrison, Paul (1987) *The Greening of Africa*, Penguin, London.
Hartmann, Jeannette (1985) "The Arusha Declaration and the Tri-Partite System of Policy Making in Tanzania" presented at the 11th Third World Conference, Chicago.
Hartnoll, A.V. (1941) "Instructions to District Officer" in the *Cory Collection*, University of Dar es Salaam.
Hartnoll, M.M. (1941) " Tribal Custom: Bahaya", T.N.A. (Bukoba District Books.)
Hafkin N.J & Bay, E.G., eds. (1976) *Women in Africa. Studies in Social and Economic Change,* Stanford.
Hawkins, H.C.G. (1965) *A Survey of Wholesale and Retail Trade in Tanganyika,* New York.
Hay M.J. and Wright, M., eds. (1982) Afican Women and the Law, Historical Perspectives, *Papers on Africa,* VII, Boston.
Hay, R. Jr. and Abu-Lughod, J. (1977) *Third World Urbanization,* Chicago.
Hedlund, Stefan and Lundahl M. (1989) *Ideology as a Determinant of Economic Systems: Nyerere and Ujamaa in Tanzania,* Scandinavian Institute of African Studies, Uppsala.
Hellberg, C.J. (1965) *Missions on a Colonial Frontier West of Lake Victoria* ,Lund.
Henderson, W.O. (1962) *Studies in German Colonial History,* London.
Heritier, Francoise (1975) "Des cauris et des hommes", in Claude Meillassoux *L'Esclavage en Afrique Pre-Coloniale,* Maspero, Paris.
Herring R.S. (1979) "Hydrology and Chronology: The Rodah Hydrometer as an Aid in Dating Interlacustrine History" in Webster, (1979.)
D'Hertefelt, M.A.A. and Trouwborst, J.H. (1962) *Les Anciens Royaumes de Zone Interlacustrine Meridionale,* Tervuren.
Hill, Polly (1986) *Development Economics on Trial: The Anthropological Case for a Prosecution*, Cambridge University Press.
Hill, Frances (1977) "Experiments with a Public Sector Peasantry" *African Studies Review* vol. 20: 25-42.
Hirji, K.F. (1979) *Colonial Ideological Apparatuses in Tanganyika under the Germans,*

Dar es Salaam.

Holm, J.D., Molutsi, P.P. and Somolekae,G. (1996) "The Development of Civil Society in a Democratic State" *ASR, African Studies Review,* vol. 39/ 2.

Huntington S. P. (1987) *Understanding Political Development: An Analytic Study,* Little, Brown, Boston.

_______ (1976) *No Easy Choice: Political Participation in Developing Countries,* Cambridge Mass.

________ (1991) *The Third Wave: Democratization in the Late Twentieth Century,* University of Oklahoma Press, Norman.

Hyden Goran (1969) *Tanu Yajenga Nchi: Political Development in Rural Tanzania,* East African Publishing House: Nairobi.

______ (1980) *Beyond Ujamaa in Tanzania: Underdevelopment and an Uncaptured Peasantry,* Heinemann, London.

_______ (1983) *No Shortcuts to Progress* University of California Press: Berkeley.

______ (1986) "The Anomaly of the African Peasantry" *Development and Change* 17 (4).

________ (1987) "Final Rejoinder" *Development and Change* 18(4).

______ (1994) "Party, State and Civil Society: Control Versus Openness" in Barkan *op cit.*

Ibrahim J. (1986) "The Political Debate and the Struggle for Democracy in Nigeria" *Review of African Political Economy* vol. 37.

Iliffe, John. (1969) *Tanganyika Under German Rule 1905-1912*, Cambridge.

_____ (1971) *Agricultural Change in Modern Tanganyika,* Nairobi.

_____ (1972)"Tanzania Under British and German Rule" in Cliffe, L. and Saul, J. *Socialism in Tanzania* Vol. 1, Nairobi.

____ (1979) *A Modern History of Tanganyika,* Cambridge U.P.

Ingham,K. (1965) *A History of East Africa,* London.

I.B.R.D.(1961) International Bank for Reconstruction and Development (World Bank) *The Economic Development of Tanganyika*, Baltimore.

James, R.W. (1971) *Land Tenure and Policy in Tanzania,* Toronto.

James, R.W. and Fimbo, G.M. (1967) *Customary Land Law of Tanzania*: A Source Book, mimeo, Faculty of Law, U.D.S.M.

Japhet, K. and Seaton, E. (1967) *The Meru Land Case,* Nairobi.

Jarvis T.S. (1937) *The Marketing of Coffee* TNA.

______ (1938) *Bukoba Coffee Inspection and Grading.* TNA.

_____ (1939) "A History of Robusta Coffee in Bukoba" *Tanganyika Notes and Records* No.8.

____ (1942) "Bukoba Coffee: Inspection and Grading" *Bukoba District Book*, Bukoba.

_____ (1950) "Crops of Bukoba District", T.N.A.

Joseph, R. (1992) "The Rebirth of Political Freedom" *Journal of Democracy* vol. 2/ 4.

Kabwato, Ernest (1976) *Causes of Decline in Major Crop Production in Bukoba: Banana, Coffee and Tea* M.A. Dissertation UDSM.

Kamuzora C. L. (1980) "Constraints to Labour Time Availability in African Smallholder

Agriculture: The Case of Bukoba District Tanzania" *Development and Change* 11(1) pp. 123-135.

_____ (1984) "High Fertility and the Demand for Labour in Peasant Economies: The Case of Bukoba District, Tanzania" *Development and Change* 15(1) pp. 105-124.

Kaniki,M.H.Y. (1979) *Tanzania Under Colonial Rule,* London.

Kasembe J.N.R., Semoka, & J.K. Samki,J.K., *Organic Farming in Tanzania,* Proceeding of the Workshop on Resource Efficient Farming, Rodale Press.

Kasfir Nelson (1986) "Are African Peasants Self-Sufficient?" Development and Change 17 (2).

Katoke, I.K. (1971) *A History of Karagwe: Northwestern Tanzania From C. 1400-1915 A.D.,* Ph.D.Dissertation, Boston University.

Katondo J. (1985) "The Theories of Centripetal and Centrifugal Migrations: The Case of Kagera Region" presented at the Third Colloquium of the Ancient Civilization of the Great Lakes Region, Makerere University.

Ketiku, A.O. & Scott-Emuakpor, M.M., (1975) *The Nutrient Content of Plantain and Banana Beverages I*badan; FAO/WHO.

Kibira, Josiah (1974) *Church, Clan and the World,* Uppsala.

Kjekshus Helge (1977) *Ecology Control and Economic Development in East African History* Heinemann, London.

Korit-Schoner, H. (1942) "Uhaya Under Clan Rule Before the Coming of the Hinda" *Tanganyika Notes and Records*, No. 14.

Krug C.A. & De Poerck R.A. (1968) *World Coffee Survey* ,Rome: FAO Agricultural Studies No.76.

Kuhlman, Tom (1994) "Organized Versus Spontaneous Settlement of Refugees" in Adelman and Sorenenson, *op cit* .

Lane, R.V. (1950) "Coffee" in Matheson, J.K. and E.W. Bovill *op cit.*

Lappé, F.M. (1971) *Diet for a Small Planet,* New York.

Larsson, Birgitta (1991) *Conversion to Greater Freedom? Women, Church and Social Change in North-Western Tanzania under Colonial Rule,* Doctoral dissertation at Uppsala University.

Laurie, W. & H. Trant (1954) *A Health Survey of Bukoba District,* East African Medical Survey Monograph no. 2, Nairobi.

Lawrence, Peter (1986) *World Recession and the Food Crisis in Africa*, James Currey, London.

Lawyers Committee for Human Rights (1995) *African Exodus,* New York.

Lemelle, S.J. (1979)"Class Struggles in Pre-Colonial Africa "*Maji Maji,* No 37, Dar es Salaam.

Lele Uma, (1984) "Tanzania: Phoenix or Icarus?" in Arnold C. Harburger, ed., *World Economic Growth: Case Studies of Developed and Developing Nations,* San Francisco, ICS Press.

Lemarchand, R. (1991) "The Political Economy of Informal Economies" in *African Insights,* vol. 21/ 4.

______ (1992) "Uncivil States and Civil Societies" *Journal of Modern African Studies* vol. 30/ 3.

______ (1995) "Rwanda: The Rationality of Genocide" in *Issue: A Journal of Opinion* vol. 23/ 2.

Leubuscher, Charlotte (1939) "Marketing Schemes" *Africa*, 12(2).

______ (1944) *Tanganyika Territory: A Study of Economical Policy under Mandate*, London.

______ (1956) *Bulk Buying from the Colonies*, Oxford University Press.

Leverett, C.W. (1957) "An Outline of the History of Railways in Tanganyika 1890-1956", T.N.R., Nos.47 and 48.

Lipset S. M. (1963) *Political Man*, Anchor, Garden City NY.

______ , Seong, K. R. and Torres,J.C. (1990) "A Comparative Analysis of the Social Requisites of Democracy," Paper presented at the Annual Convention of the American Political Science Association , San Francisco, August 30- September 3.

Lipton, Michael (1968) "The Theory of the Optimising Peasant" *Journal of Development Studies* vol. 4: 327-51.

Lofchie, Michael F. (1976) "Agrarian Socialism in the Third World: The Tanzanian Case" *Comparative Politics* vol.8.

Lugard, F.D. (1893) *The Rise of Our African Empire* 2 Vols., Beachwood.

Lyall, A.B. (1973) *Land Law and Policy in Tanganyika 1919-1932*, L.L.M. Thesis, U.D.S.M.

Mackenzie, Fiona (1988) "Reconceptualizing Land Tenure Systems: Murang'a District Kenya" Paper prepared for the Annual Conference of the Canadian Association of African Studies, Queens University, Kingston, 11-14 May.

Mair, L.P. (1934) *An African People in the Twentieth Century*, London.

______ *African Kingdoms*, Oxford.

Maliyamkono, T.L. and. Bagachwa, *M.S. D. The Second Economy in Tanzania* , James Currey, London.

Mamdani, M. (1976) *Politics and Class Formation in Uganda*, London.

Maquet, J.J. (1961) "Une Hypothèse pour l'Étude des Féodalités Africaines" *Cahiers d"Études Africaines*, No. 6.

Mangat, J.S. (1969) *The History of the Asians in East Africa*, London.

______ (1971) *Power and Society in Africa*, London.

Mascarenhas, Ophelia, and Mbilinyi, M., (1983) *Women in Tanzania, an Analytical Bibliography* Scandinavian Institute of African Studies, Uppsala.

Matheson J.K. & Bovill E.G. (1950) *East African Agriculture*, London.

Mbilinyi Marjorie ed. (1985) *Cooperation or Exploitation*, ILO, Geneva.

Mbilinyi, Simon (1976) *Coffee Production in Tanzania*, Nairobi.

McCall Michael (1987) "Carrying Heavier Burdens but Carrying Less Weight: Some Implications of Villagization in Tanzania" in J.H. Momsen & J.G. Townsend eds. *Geography of Gender in the Third World*, State University of New York Press, Hutchinson.

McCarthy D.M.P. (1982) *Colonial Bureaucracy and Creating Underdevelopment: Tanganyika,1919-1940,* Ames: Iowa State University Press.

McMaster D.N. (1960) "Change of Regional Balance in the Bukoba District of Tanganyika" *The Geographical Review* 50 (1).

McHenry, Dean (1979) *Tanzania's Ujamaa Villages: The Implementation of a Rural Development Strategy* Research Series no. 39, Institute of International Studies, Berkeley.

Meeker, J, and Meekers D. (1997) "The Precarious Socio-Economic Position of Women in Rural Africa: The Case of the Kaguru of Tanzania" in *African Studies Review* vol. 40/ 1.

Meillassoux, C. (ed) (1971) *The Development of Indigenous Trade and Markets in West Africa,* Oxford.

_____ (1981) *Maidens, Meal and Money: Capitalism and the Domestic Community,* Cambridge.

Migot-Adholla S.E. (1979) "Rural Development Policy and Equality" in *Politics and Public Policy in Kenya and Tanzania,* eds. Joel D. Barkan and John Okumu Praeger Publishers, New York.

Mihyo N.Z. (1975) *Peasant Production and Differentiation in Kabirizi Division,* unpublished Masters thesis University of Dar es Salaam.

Mihyo, P.B. (1981) *Industrial Conflict and Change in Tanzania,* Tanzania Publishing House, Dar es Salaam.

Mikell, G. (1985) "Expansion and Contraction in Economic Access for Rural Women in Ghana" *Rural Africana* vol.2, pp.113-130.

Milne, G. (1938) "Bukoba: High and Low Fertility on a Laterised Soil" *The East African Agricultural Journal.*

Mishambi, G.T. (1978) *Peasantry Under Imperialism: A Case Study of West Lake Region Tanzania,* M.A. Thesis, U.D.S.M.

Moller, Mattis (1958) "Bahaya Customs and Belief in Connection with Pregnancy and Childbirth" *Tanganyika Notes and Records* no. 50.

Montague, F.A. and F.H. Page-Jones, (1951) "Some Difficulties in the Democratization of Native Authorities in Tanganyika" *Journal of African Administration,* Vol. 3, No. 1.

Moore, B. Jr. (1966) *Social Origins of Dictatorship and Democracy,* Boston.

Morris-Hale, W. (1969) *British Administration in Tanganyika from 1920 to 1945,* Thèse No. 192, Génève.

Mueller, Suzanne, D. (1980) "Barriers to Further Development of Capitalism in Tanzania The Case of Tobacco", Unpublished paper presented at Canadian Association of African Studies, Guelph, Ontario.

Mukoyogo, M.C. (1975) Nyarubanja Tenure System: Mechanics of Enfranchisement, L.L.B. Dissertation, U.D.S.M.

Munishi, G. K. (1989) "Bureaucratic Feudalism, Accountability and Development in the Third World: The Case of Tanzania" in J. G. Jabbra and O.P. Dwivedi (eds.) *Public Service Accountability,* Kumarian Press, West Hartford.

Musoke, I.K.S. (1971) "Building Socialism in Bukoba", in Proctor, I.E. *Building Ujamaa Villages in Tanzania,* Dar es Salaam.

Mutahabu, G.R. (1969) *The Importance of Peasant Consciousness for Effective Land Tenure Reform,* B.A. Thesis, U.D.S.M.

Nabudere, D.W. (1977) "The Closer Union Issue and its Implications for the Evolution of the British Colonial Policy of East Africa: 1923-1931", mimeo Seminar Paper, U.D.S.M.

_____ (1989) "The One Party State in Afrca and its Assumed Philosophical Roots" in P. Meyns and D. W. Nabudere (eds.) *Democracy and the One-Party State in Africa,* Institut fur Afrika-Kund, Hamburg.

Newbury C. (1995) "Background to Genocide in Rwanda" *Issue: A Journal of Opinion* vol. 23/ 2.

Ngaiza M.K. (1981) *Some Aspects of Food Production and Food Wastages in Muleba District, Kagera Region: The Case of Non-Grain Food Production and the Utilization of Post Harvest Technology*, unpublished Masters Thesis, The University of Dar es Salaam.

Nindi, B. (1978) *Agricultural Change and Rural Class Formation in Iringa District Tanzania,* Ph.D. Thesis, Hull University.

Ninsin, K. (1988) "Three Levels of State Reordering: The Structural Aspects" in D. Rothchild and N. Chazan (eds.) *Precarious Balance: State and Society in Africa,* Westview, Boulder.

Northcote R.C. (1936) *Report on Bukoba Coffee Marketing,* TNA 215/1410.

Noumoff, S. J. (2000) "Civil society: Does it have Meaning?" A paper presented at the Centre for Developing Area Studies, McGill University, September 21st.

Nyang'oro, J. E. (1989) *The State and Capitalist Development in Africa: Declining Political Economies,* Praeger, New York.

_____ (1992) "Building Pluralist Democracy in Africa" in *Midsouth Political Science Journal* vol. 13.

_______ (1994) "Reform Politics and the Democratization Process in Africa" *African Studies Review* vol. 37/ 1.

Nyerere, Julius K. (1977) *The Arusha Declaration Ten Years After,* Government Printer, Dar es Salaam, Tanzania.

Obbo, Christine (1980) *African Women: Their Struggle for Economic Independence,* London.

Oberg, K. (1940)"The Kingdom of the Ankole in Uganda" in Fortes, M. and Evans-Pritchard, E. *African Political Systems,* Oxford.

O' Donnell, G. (1978) "Reflections on the Pattern of Change in the Bureaucratic-Authoritarian State" in *Latin American Review* vol. 8.

____, and Schmitter, P. (1986) *Transitions from Authoritarian Rule: Tentative Conclusions about Uncertain Democracies,* Johns Hopkins University Press, Baltimore.

Ogot, B.A. (ed) (1968) *Zamani: A Survey of East African History*, Nairobi.

Oliver R. (1952) *The Missionary Factor in East Africa,* London.

O'Neill, Norman and Mustafa, K., editors, (1990) *Capitalism, Socialism and the Development Crisis in Tanzania* Avebury, Aldershot.

Orde-Browne, G.J. (1926) *Report Upon Labour in the Tanganyika Territory*, London.

_____ (1946) *Labour Conditions in East Africa*, London.

Ortner, S. B. (1984) "Theory in Anthropology since the Sixties" in *Comparative Studies in Society and History* vol. 26/ 1.

Palma, G. (1978) "Dependency: A Formal Theory of Underdevelopment or a Methodology for the Analysis of Concrete Situations of Underdevelopment", *World Development,* Vol 6.

Palmer R. and N. Parsons (1977) *The Roots of Rural Poverty in Central and Southern Africa*, University of California Press, Berkeley.

Phillipson, D.W. (1977) *The Later Prehistory of Eastern and Southern Africa*, London.

Pokorny, Dusan (1973) "The Haya and Their Land Tenures: Property Rights and the Surplus Problem" *Rural Africana* No.22.

Posnansky, M. (1975) "Connections Between the Lacustrine Peoples and the Coast", in Chittick and Rothberg, *op cit*.

Post, K. (1991) "The State civil Society, and Democracy in Africa" in R. Cohen and H. Gouldbourne (eds.) *Democracy and Socialism in Africa,* Westview, Boulder.

Potash, Betty ed. (1986) *Widows in African Societies: Choices and Constraints*, Stanford University Press.

Powesland, P.Q. (1951) "History of the Migration in Uganda", in Richards A.I. ed. *op cit*.

Putnam, Robert. (1995) "Bowling Alone" *Journal of Democracy,* January.

Raikes, Philip (1976) *Coffee Production in West Lake Region Tanzania,* Project Paper A.76.9, Institute for Development Research Copenhagen.

_____(1978) "Rural Differentiation and Class Formation in Tanzania", *Journal of Peasant Studies* Vol.5, No 3.

_______ (1986) "Eating the carrot and wielding the stick: the agricultural sector in Tanzania" in Boesen J., *et al* eds. *op cit*.

Rald J. & Rald K. (1975) *Rural Organization in Bukoba District Tanzania* Uppsala: SIAS.

Ramm-Ericson (1966)"Price Movements of East Africa's Exports" in Smith, H.E. (ed) *op cit.*

Reeve, H. (ed) (1903) *The Grenville Memoirs,* Vol.6.

Reining, Priscilla (1962) "Haya Land Tenure: Landholding and Tenancy" *Anthropological Quarterly* 35 (1).

_______ (1967) *The Haya, Agrarian System of Sedentary People,* Ph.D. dissertation, University of Chicago.

_____ (1970) "Social Factors and Food Production in an East African Peasant Society: The Haya" in S.M. Peter, ed. *African Food Production Systems* Macloughlin, Baltimore.

Rennie, J.K. (1972) "The Pre-Colonial Kingdom of Rwanda: A Reinterpretation" *Transafrican Journal of History,* Vol. 2,No.2.

Resnick Adrian N. (1981) *The Long Transition: Building Socialism in Tanzania*, Monthly Review Press, New York.

Richards, A.I. (1951)"The Travel Routes and the Travellers" in Richards, A.I. (ed) *op cit.*

_____ (ed)(1951b) *Economic Development and Tribal Change: A Study of Immigrant Labour in Buganda,* Cambridge.

Richards A.I. and Reining Priscilla (1954) "Report on Fertility Surveys in Buganda and Buhaya" in Lorimer, F. ed. UNESCO.

Roberts, A. (ed.) (1968) *Tanzania Before 1900: Seven Area Histories,* Nairobi.

Robertson C. & I. Berger (1986) *Women and Class in Africa,* Africana Publishing Company, New York.

Robinson, P. T. (1994 "Democratization: Understanding the Relationship between Regime Change and the Culture of Politics" in *African Studies Review* vol. 37/1.

Rodemann, H.Q. (1961) *Tanganyika 1890-1940: Selected Aspects of German Administration* Ph.D. Dissertation, University of Chicago.

Rodney, W. (1972) *How Europe Underdeveloped Africa,* London.

_____ (1979) "The Political Economy of Colonial Tanganyika 1830-1930" in Kaniki *op cit.*

Roitman, J. (1990) "The Politics of Informal Markets," in *Jornal of Modern African Studies* vol. 20/ 4.

Rostow, W.W. (1971) *Politics and the Stages of Growth,* Cambridge University Press.

_____ (1990) *Theories of Economic Growth from David Hume to the Present,* Oxford University Press.

Rustow, D. A. (1970) "Transitions to Democracy: Toward a Dynamic Model," in *Comparative Politics* vol. 2/ 3.

Rweyemamu, J. (1973)*Underdevelopment and Industrialization in Tanzania,* Nairobi.

Sandbrook, Richard (1988) "Liberal Democracy in Africa: A Socialist Revisionist Perspective" *Canadian Journal of African Studies* vol. 22/ 2.

____ (1993) *The Politics of Africa's Economic Recovery,* Cambridge University Press.

_____ (1995) "Bringing Politics Back In? The World Bank and Adjustment in Africa" *Canadian Journal of African Studies* vol. 29/ 2.

Saul, J. and Woods, R. (1971)"African Peasantries" in Shanin, T.(ed) *Peasants and Peasant Societies,* Harmondsworth.

Saul, Mahir (1981) "Beer, Sorghum and Women: Production for the Market in Rural Upper Volta" *Africa: Journal of the International African Institute*, London pp.747-764.

Savage, Stephen P. (1977) "Talcott Parsons and the Structural-Functionalist Theory of the Economy" in B. Hindess *Sociological Theories of the Economy,* Macmillan.

Schatzberg, M., G. (1993) "Power, Legitimacy and 'Democratisation' in Africa" in *Africa* vol. 63/ 4.

Schmidt, P.R. (1974) *An Investigation of Early and LateIron Age Culture Through Oral Traditions and Archaeology: AnInterdisciplinary Case Study in Buhaya, Tanzania*, Ph.D. Thesis, Northwestern University.

______ (1979) *Historical Archaeology* Greenwood Press.

Schumacher E.F. (1973) *Small is Beautiful* Abacus, London.

Sheriff, A.M.H. (1979) "Tanzania Societies at the Time of the Partition" in Kaniki, *op cit.*

Shivji, I., G. (1969) *Land Tenure and Agricultural Development in Kenya and Tanzania,* Unpublished manuscript, U.D.S.M.

______ (1975a) *Class Struggles in Tanzania*, London.

_____(1975b) "Peasants and Class Alliances", *Review of African Political Economy,* No.3.

_____(1991) "Contracictory Class Perspectives in the Debate on Democracy" in Shivji I., G. (ed) *State and Constitutionalism an African Debate on Democracy* SAPES, Harare.

Shoemaker, R., (19950 "Assessing Democratic Governance: Expanding Civil Society and Participation in the Political Process" in *African Voices: A Newsletter on Democracy and Governance in Africa* USAID vol. 4/ 1.

Sisulu, E. with Immam, A., and Diouf, M., (1991) "Report on the Workshop on Gender Analysis and African Social Science" *CODESRIA Bulletin* number 4.

Slater, D. (1974) " Colonialism and the Spatial Structure of Underdevelopment: Outlines of an Alternative Approach with Special Reference to Tanzania", in *Progress in Planning*, Vol.4, Part II, London.

Smelser, N., (1964) "Toward a Theory of Modernization" in A. Etzioni and E. Etzioni (eds.) *Social Change*, Basic Books, New York.

Smith C.D. (1985) *The Commoditization of African Agriculture and Changed Social Relations:The Case of the Haya of Tanzania*, Unpublished doctoral dissertation, University of Essex, Sociology.

____(1987) "Smallholder Farming in Kagera Region, Tanzania: Constraints to Coffee Production", *Labour, Capital and Society* 20 (2).

_____ (1988) "Farming and Income Generation in the Female-Headed Smallholder Household: The Case of a Haya Village in Tanzania" co-authored by Lesley Stevens, in *the Canadian Journal of African Studies* Volume 22 Number 3.

______ (1989) *Did Colonialism Capture the Peasantry? A Case Study of the Kagera District Tanzania?* Research Report No. 83 published by the (Scandinavian Institute of African Studies), Uppsala, Sweden.

______ (1990) "Peasant Resistance to the Grow More Coffee Campaign in Kagera Region Tanzania" in *Canadian Journal of Development Studies* Volume 10 Number 2.

Smith, H.E. (ed) (1967) *Readings on Economic Development and Administration in Tanzania,* London.

So A.Y. (1990) *Social Change and Development* Sage Publications, Newbury Park.

Sorrenson, M.P.K. (1968) *Origins of European Settlement in Kenya,* Nairobi.

Southall, A.W. and Gutkind P., (1957) *Townsmen in the Making: Kampala and its Suburbs,* Kampala, East African Institute of Social Research.

Special Correspondent (1929) "Indirect Rule in Tanganyika" in *East Africa,* Vol.5, No

23.

Stamp P., (1991) "Burying Otieno: Politics of gender and Ethnicity in Kenya" in *Signs* vol. 16/ 4.

Staudt, Kathleen (1982) "Women Farmers and Inequities in Agricultural Services", in E. Bay, Women and Work in Africa, *op cit.*

Stevens, Lesley (1991) "Religious Change in a Tanzanian Village" *Journal of Religion in Africa*, E.J. Brill.

Storgaard, Birgit (1976) "Women in ujamaa villages" *Rural Africana* 29: 135-155.

Stuart, M. (1945) "Review of Commercial Conditions" *East Africa Pamphlet*, No.336.

Sundkler, Bengt. (1980) *Bara Bukoba, Church and Community in Tanzania,* Uppsala.

Swantz, Marja-Liisa (1977) "Strain and Strength among Peasant Women in Tanzania". University of Dar es Salaam, BRALUP Research Paper 49.

______ (1985) *Women in Development: A Creative Role Denied?*, London, C. Hurst Co.

Tandon, Y.A. (1973) *Problems of a Displaced Minority: The New Position of East Africa's Asians,* London.

Tanganyika Gazette

Tanganyika Provincial Commissioners (1935-1959) *Annual Reports,* Dar es Salaam.

Tanganyika Territory (1927) "Native Administration", *East Africa Pamphlet*, No.91.

Tanganyika Territory (1937) "Notes on Cultivation Preparation and Marketing of Arabica Coffee", *East Africa Pamphlet,* No. 175.

Tanganyika Territory (1939-1945) *Trade and Information Reports.*

Temu, A.J. (1979) "Tanzanian Societies and Colonial Invasion 1875-1907" in Kaniki *op cit.*

Terray, E. (1972) *Marxism and Primitive Societies,* New York and London.

______ (1974) "Long Distance Trade and the Formation of the State, the Case of the Abron Kingdom in Gyamon" in *Economy and Society,* Vol.3,No.3.

Tibaijuka A.K. (1979) *Strategies for Smallholder Agricultural Development in West Lake Region Tanzania,* Uppsala, Swedish University of Agricultural Sciences, Report No. 150.

______ (1984) *An Economic Analysis of Smallholder Banana-Coffee Farms in the Kagera Region,Tanzania*, Uppsala: Swedish University of Agricultural Sciences.

______ (1985) *Factors Influencing The Cultivation of Firewood Trees on Peasant Farms*, Uppsala, Swedish University of Agricultural Sciences.

TANGANYIKA PROVINCIAL COMMISSIONERS (1935-1959) *Annual Reports* London: HMSO.

Timberlake, Lloyd (1986) *Africa in Crisis,* Philadelphia, Earthscan.

Thomas, C. (1974) *Dependence and Transformation*, New York.

Thorner, D. (1971)"Peasant Economy as a Category in Economic History" in Shanin, T. *op cit.*

T.N.A. *The National Archives of Tanzania* various years
Bukoba Blue Books, 1931-1948.

District Book for Bukoba (1928).
Primitive Native Eduction System, in Bahaya.
Laws Manners Customs Sheets (1929a), (1929b).
Bukoba District Book "Bahaya Land Tenure: Bukoba" (1929C).
File 180/B, "Land Tenure and Tribute in Bihamamulo" (1929d).
"Nyarubanja, in the Bukoba District Historical Notes" (1936).
"Notes on the Cultivation, Preparation and Marketing of Coffee" (1937).
Davies, A.H., "Rough Classification of Bananas" (1948).
District Book for Bukoba, "Financial Statements of Bukoba Central Native Treasury.
File 619, Registration No.969 (*Nyarubanja* Registers).
File 446, Nyarubanja Rules.
File 619, October (1955) *"Petition Against Nyarubanja Tenure Plus Further Evidence Against Nyarubanja Tenure"*.
File 619, Vol. IV (1959), Letters of H.J. Cook, District Commissioner.
File 622 (On land ownership).
215/1410 (1936), R.C. Northcote, *Report on Bukoba Coffee Marketing*.
215/50, McMillen to Bakama, 4 February 1920.

Todaro M. P. (1976) *Internal Migration in Developing Countries* International Labour Office, Geneva.

Tothill, J.D. (1940) *Agriculture in Uganda,* London.

Tripp A., (1994) "Gender, Political Participation and Associational Life in Uganda and Tanzania" in *African Studies Review* vol. 37/ 1.

Uziogwe, U.N. (1968) "Pre-Colonial Markets in Bunyoro-Kitara" in Ogot, B.A. *op cit.*

Vidal, C. (1974) "Economie de la societé féodale rwandaise" *Cahiers d'Études Africaines.*

Von Freyhold, M. (1974) "The Rise and Fall of Colonial Modes of Production", *Institute of Finance Management Paper,* Dar es Salaam.

_____ (1977) The Formation of a Class of Wage-Laborers in Tanzania Under Colonial Capitalism", mimeo History Dept. Seminar Paper, U.D.S.M.

_______ (1979) *Ujamaa Villages in Tanzania: Analysis of a Social Experiment,* New York, Heinemann.

Wallerstein, I. (1970)"The Colonial Era in Africa: Changes in the Social Structure" in Gann, C. and Dugan, P. (eds.) *Colonialism in Africa 1870-1960,* Vol. 2, Cambridge.

______ (1976) "The Three Stages of African Involvement in the World Economy" in Gutkind, P.C. and Wallerstein, I. (eds) *The Political Economy of Contemporary Africa,* Beverley Hills/London.

Warren, Bill (1973) "Imperialism and Capitalist Industrialization" New Left Review, No. 81.

_____ (1980) *Imperialism Pioneer of Capitalism,* London.

Waters A.R. (1972) "Change and Evolution in the Structure of the Kenya Coffee Industry" African Affairs vol.71 no.283.

Webster J.B. (1979) *Chronology, Migration and Drought in Interlacustrine Africa,* New York.

____ (1979a) Noi! Noi! Famine as an aid to Interlacustrine Chronology" in Webster, J.B. *op cit.*

Were, G. M. (1987) "The Western Bantu Peoples from A.D. 1300-1800" in Ogot, B.A. (ed) *op cit.*

White, Luise (1980) *Women's Domestic Labor in Colonial Kenya Prostitution in Nairobi* African Studies Center, Boston. Working Papers no. 30.

________ (1990) *The Comforts of Home: Prostitution in Colonial Nairobi*, Chicago: Universtiy of Chicago Press.

Whitely, W.H. (ed) (1971)_*Baisha ya Hamed Bin Muhammed El Murjebi Yani Tippu Tip,* Nairobi.

Williams Gavin (1987) "Primitive Accumulation: The Way to Progress" *Development and Change* 18 (2).

Wolcott Harry F. (1974) *The African Beer Gardens of Bulawayo,* New Brunswick N.J.: Rutgers Center of Alcohol Studies.

Wood D., (1992) "Civil Society in Europe and Africa: Limiting State Power Through the Public Sphere" *African Studues Review* vol.35/2.

World Health Organization WHO (1980) "Problems Related to Alcohol Consumption" *Technical Report Series* #650 Geneva; WHO.

World Bank (1981) *Accelerated Development in Sub-Saharan Africa: An Agenda for Action* Washington D. C.

______ (1989) *Sub-Saharan Afrcia: From Crisis to Sustainable Growth*, Washington D.C.

Wright, Marcia (1976) *Educational Policies Between the Wars in Tanganyika*, University of Dar es Salaam.

Wrigley, C.C.(1978) *Crops and Wealth in Uganda,* Kampala.

Wyatt, A.W. (1931) "Tribal History and Legend: Bahaya", T.N.A.

Young, Crawford (1994) "Democratization in Africa: The Contradictions of a Political Imperative" in J. Widner (ed) *Economic Change and Political Liberalization in Africa,* Johns Hopkins University Press, Baltimore.

Yotopoulos, P.A. Lau. L. and Kutlu S. (1970) "Labour Intensity and Relative Efficiency in Indian Agriculture" *Food Research Institute Studies, Vol.9,No.1.*